TECHNOLOGIES OF SPECULATION

Technologies of Speculation

The Limits of Knowledge in a
Data-Driven Society

Sun-ha Hong

NEW YORK UNIVERSITY PRESS
New York

NEW YORK UNIVERSITY PRESS
New York
www.nyupress.org

References to Internet websites (URLs) were accurate at the time of writing. Neither the author nor New York University Press is responsible for URLs that may have expired or changed since the manuscript was prepared.

Library of Congress Cataloging-in-Publication Data
Names: Hong, Sun-ha, author.
Title: Technologies of speculation : the limits of knowledge in a data-driven society / Sun-ha Hong.
Description: New York : New York University Press, 2020. |
Includes bibliographical references and index.
Identifiers: LCCN 2019041465 | ISBN 9781479860234 (cloth) |
ISBN 9781479883066 (paperback) | ISBN 9781479855759 (ebook) |
ISBN 9781479802104 (ebook)
Subjects: LCSH: Technology—Social aspects. | Artificial intelligence. | Algorithms.
Classification: LCC T14.5 .H636 2020 | DDC 303.48/3—dc23
LC record available at https://lccn.loc.gov/2019041465

New York University Press books are printed on acid-free paper, and their binding materials are chosen for strength and durability. We strive to use environmentally responsible suppliers and materials to the greatest extent possible in publishing our books.

Manufactured in the United States of America

10 9 8 7 6 5 4 3 2 1

Also available as an ebook

CONTENTS

Introduction

Driven by Data

What counts as knowledge in the age of big data and smart machines?

We—we as that fiction of a collective public, we as individuals cut to ever finer pieces with each measurement—are becoming, like it or not, "data-driven." Externally, smart machines and algorithmic prediction take the wheel, knitting together an expansive landscape of facts against which individuals are identified and judged. Internally, human drives—which, Deleuze understood, are not merely internal psycho-phenomena but themselves social structures[1]—are measured and modulated through ubiquitous sensors. The rapid expansion of technologies of datafication is transforming what counts as known, probable, certain, and in the process, rewriting the conditions of social existence for the human subject.

The data-driven society is being built on the familiar modern promise of better knowledge: data, raw data, handled by impartial machines, will reveal the secret correlations that govern our bodies and the social world. But what happens when the data isn't enough and the technology isn't sufficient? The limits of data-driven knowledge lie not at the bleeding edge of technoscience but among partially deployed systems, the unintended consequences of algorithms, and the human discretion and labor that greases the wheels of even the smartest machine. These practical limits provoke an array of speculative practices, putting uncertainties to work in the name of technological objectivity. Weak indicators of human behavior and other fragmentary, error-prone data are repackaged into probabilistic "insights," whose often black-boxed deployment now constitutes a global industry. Futuristic imaginaries of posthuman augmentation and absolute predictivity endow today's imperfect machines with a sense of legitimacy. In the process, technologies of datafication are reshaping what *counts* as knowledge in their own image. From self-surveillance to counterterrorism intelligence, the business of datafication hinges on redefining what kinds of data in whose hands should determine the truth of who I am or what is good for me.

The moral and political question, then, is not simply whether datafication delivers better knowledge but how it transforms what counts in our society: what counts for one's guilt and innocence, as grounds for suspicion and surveillance, as standards for health and happiness. Datafication thus reprises the enduring dilemma around the modern ideal of the good liberal subject: individuals who think and know for themselves, their exercise of reason founded on free access to information and the possibility of processing it fairly. New technologies for automated surveillance and prediction neither simply augment human reason nor replace it with its machinic counterpart. Rather, they affect the underlying conditions for producing, validating, and accessing knowledge and modifying the rules of the game of how we know and what we can be expected to know. The promise of better knowledge through data depends on a crucial asymmetry: technological systems become increasingly too massive and too opaque for human scrutiny, even as the liberal subject is asked to become increasingly legible to machines for capture and calculation.

The Duality of Fabrications

These dilemmas show that when big data and smart machines produce new predictions, new insights, what they are creating are fabrications: a process by which approximations are solidified into working certainty, guesswork is endowed with authority, and specific databases and algorithms—and all the biases and heuristics they embody—are invested with a credibility that often outstrips their present achievements. To call these activities fabrications does not mean that datafication is merely a con of epic proportions. The word originates from *fabricare*, to manufacture with care and skill; a manufacturing that every kind of knowledge system, from science to religion, undertakes in its own way. To analyze datafication in this way is to understand how data is seizing and affirming its position as truth-maker today.

Often, such fabrications involve highly accurate and sophisticated measurements that tend to perform best within tightly prescribed parameters. At the same time, their application to real-world problems often relies on arbitrary classifications, messy data, and other concealed uncertainties. Exercise trackers combine advancements in miniaturized sensors with rough heuristics, such as ten thousand steps per day—a figure originally invented by mid-twentieth century Japanese marketers to sell pedometers—to produce their recommendations. Large-scale systems, such as electronic surveillance systems for counterterrorism purposes, embed layers of human labor and decision-making into a process that is ultimately black-boxed to the ordinary citizen. The connections between data, machines and "better knowledge" remain obscure for most of us, most of the time. In many concrete cases, the claim to better, more objective knowledge through data also depends on shifting expectations around what looks and sounds like reliable truth.

Fabrications are therefore ambiguous and unstable things. Imperfect algorithms, messy data, and unprovable predictions are constantly intersected with aspirational visions, human judgment, and a liberal dose of black-boxing. Importantly, such a duality is normal to the work of datafication: a feature, not a bug. Accordingly, the solution is not as simple as a bug report that sorts out the good kinds of data-driven knowledge from the bad. Such clean and neat distinctions are not always possible and risk supporting the technocentric imagination that a few rounds of bug-fixing would allow the data to truly provide better knowledge anywhere and everywhere. Instead, this book traces underlying patterns in how such claims are made—an approach that has been crafted across areas such as sociology of knowledge, history of science and technology, and critical data studies.[2] The manufacturers and distributors of data-driven fabrication do not simply "cheat" truth. Rather, they are playing the game of making certain kinds of truth count. What emerges is not so much a whole new regime of knowledge but new opportunities for extending, distorting, and modifying long-standing tendencies for how we use numbers and machines to make sense of our worlds.

This approach also situates the technologies of our time in the long history of data, quantification, and social sorting. As buzzwords of the day, big data or smart machines have a short and specific life span (even if they, like artificial intelligence, often end up being recycled). But the underlying shift in what counts as knowledge often outlasts those moments in the spotlight. Joseph Weizenbaum, a pioneer of AI, had identified this dynamic in an earlier generation of computing technologies: that far before and far in excess of computers being shaped to serve humans, humans are asked to become more compatible with the machines themselves.[3] From the human body pressed into mechanical action in the Fordist factory, as immortalized in Charlie Chaplin's *Modern Times*, or the twenty-first-century population of "ghost workers"[4] performing invisible, low-paid labor to support AI systems, the sublime spectacle of computing power constantly relies on a scaffolding of machine-compatible humans. From an epistemological standpoint, the fabrications captured in this book also echo the social life of earlier technologies for datafying bodies and lives, where the gradual normalization of modern attitudes toward numbers and statistics, then machine-driven databases, as objective fact was often achieved for and through specific political exigencies of the day.[5]

Similarly, today's fabrications are thoroughly imperfect and inescapably political. Insofar as the data-driven society is built on the bullish promise of a world run more rationally and objectively, this optimism feeds off contemporary anxieties about the seemingly growing uncertainties of modern life. There is the global diffusion of micro-threats in the "war on terror," emblematized by the almost random possibility of a "lone wolf" attack, or the heightened pressure for citizens to optimize their everyday life routines to survive the neoliberal market. Yet even as uncertainty functions as the bogeyman Other to the seductive promises of datafication, such knowledge is often achieved precisely by putting uncertainties to work. In the gaps between the fantastic promises of technology and its imperfect applications, between the reams of machine-churned knowledge and the human (in)ability to grasp it, grows a host of emergent, speculative practices that depend on the twisted symbiosis of knowledge and uncertainty.

Out There, In Here

This book examines two sites where datafication is turning bodies into facts: shaping human life, desire, and affect into calculable and predictable forms and, in doing so, changing what counts as the truth about those bodies in the first place. The first is the Snowden affair and the public controversy around the American government's electronic "dragnet"[6] surveillance technologies, built to quietly collect phone, email, and other electronic communications data at an unprecedented scale. The second is the rise of miniature, automated tracking devices for the monitoring of everyday life, from exercise levels to emotional status, and the subsequent analysis of that data for personalized "insights." Surveillance by the state, surveillance by oneself—these practices reflect the expanding reach of big data's rationality across established boundaries of the public and the private, the political and the personal.

On December 1, 2012, one "Cincinnatus"—namesake of that mythical Roman embodiment of civic virtue—contacted journalist Glenn Greenwald to request an encrypted conversation. He received no reply. Six months later, Cincinnatus was revealed to be Edward Snowden, formerly a subcontractor for the National Security Agency, now a fugitive wanted by the United States. Having eventually succeeded in reaching Greenwald, he enlisted the journalist's help in leaking a massive cache of classified information, revealing a sprawling range of high-tech surveillance programs wielded by the US and other Western governments.

Somewhere on the way, a philosophical question had emerged: What can the public know, and what is made the public's duty to know? The programs Snowden publicized entailed the collection of personal communications data in enormous quantities through methods designed to remain totally imperceptible to the population subject to it. This data would be harnessed toward predictive calculations whose efficacy often cannot be publicly (and, sometimes, even privately) proved. As the leaks fanned an international controversy starring lawsuits and policy debates, award-winning documentaries, and presidential speeches, the public was caught in uncertainty. One letter to *The New York Times* read: "What kind of opinion can a citizen have of his government when his government is unknown to him, or, worse, is unknown to itself? After 9/11, we found ourselves in a state of war with faceless terrorists . . . but those we have empowered to protect us use methods that we cannot see, taste or smell."[7] Popular book titles spoke of shadow governments, dragnet nations, and no place to hide.[8] Such metaphors spoke to a deep sense of asymmetry: How can ordinary human subjects know the world "out there," a world governed by increasingly vast and complex technological systems, a world that seems to begin where our personal experiences and lived worlds fall away?

As America and the world grappled with the implications of Snowden's leaks, similar dilemmas around knowledge and uncertainty were playing out through a very different set of fantasies around progress and empowerment. In September 2011, Ariel Garten took to the stage for a TED Talk—a series famous for providing slick, punchy briefs about the pressing problems of the day, and, more often than not, optimism that they can be overcome through technological and social innovations. Garten was well suited for such a stage. Juggling a life as a fashion designer, psychotherapist, artist, neuroscience researcher, and entrepreneur, she could present a figure of someone riding the waves of the newest technologies, someone standing at the threshold of the near future. Garten enthused about a wearable brainwave sensor on her forehead—an electroencephalography device that would soon go on sale by the name of "Muse." It will tell us how focused or relaxed we are, she said, revealing aspects of ourselves that had previously been "invisible":[9]

> My goal, quite simply, is to help people become more in tune with themselves. I take it from this little dictum, "Know thyself." . . . I'm here today to share a new way that we're working with technology to this end to get familiar with our inner self like never before—*humanising technology* [emphasis mine] and furthering that age-old quest of ours to more fully know the self.

As the American government invested massive sums into data-driven, predictive surveillance systems, its tech enthusiasts and entrepreneurs were using similar techniques to pursue an individualistic and posthuman vision: the human subject—ever a blind amnesiac, fumbling its way through the maze that is its own body and mind—would be accompanied by machines that would correct its memories and reject its excuses. Technologies of self-surveillance, overlapping across categories such as biohacking and lifelogging, use miniaturized smart machines to enable highly persistent and automated processing of human life into data-driven predictions. From the predictable, such as measures of exercise and sleep, to the bizarre, such as sex statistics (thrusts per minute), self-surveillance promises to bring home the benefits of datafication, enabling a more objective basis for knowing and improving the self.

The transformation of the everyday into a persistent backdrop of measurements and nudges promises unprecedented knowledge for the human subject precisely by shifting accepted norms around what counts as better knowledge. At one level, these machines track individuals in ways inaccessible to the human subjects' own cognition and experience—either by measuring phenomena beyond the capacity of the human senses, such as the electrical conductance in the skin, or by measuring at a frequency and granularity that people realistically cannot keep up with. The problem of what we can and must know is thus brought back from "out there" to the "in here" of the individual body and life. What does it mean to "know myself" if that knowing is achieved through mass-produced, autonomously operative devices? What kind of relationship to knowledge is produced when machines communicate ceaselessly with the body and with each other along channels that my conscious reflection cannot ever access? In many ways, the pursuit of the datafied self reenacts Weizenbaum's dictum: the capture of bodies for predictive analytics encourages those bodies to behave in ways that are most compatible with the machines around them—and, by extension, the institutions behind those machines. The good liberal subject is thus rearticulated as tech-savvy early adopters (who are willing to accept the relations of datafication before their validity is proved) and as rational, data-driven decision makers (who learn to privilege machinic sensibility above human experience).

The book traces the public presentation of state and self-surveillance across multiple sites where the technologies and their associated fantasies are proclaimed, doubted, justified, and contested. This includes the media coverage, leaked government files, lawsuits and Senate hearings around the Snowden affair (2013–), as well as advertising and promotional discourse, media coverage, and conversations with entrepreneurs and enthusiasts around the rise of self-surveillance technologies (2007–).[10] I also draw on observational fieldwork of the Quantified Self (QS), an international community of self-trackers that has played a key role in popularizing the technology from a niche geek endeavor to a market of millions of users. Despite clear differences in the specific configuration of state and corporate interests, the interpellation of citizens and consumers, certain ways of thinking and dreaming about datafication recur across these contexts. Chapter 1 lays out the technological fantasies that help justify, make sense of, and lend excitement to concrete systems of data-driven truth making. The promise of better knowledge is here broken down into a historically recurring faith in technoscientific objectivity, through which datafication promises a certain epistemic purity: a raw and untampered representation of empirical reality, on which basis human bodies and social problems might also be cleansed of complexity and uncertainty. These fantasies serve as navigational devices for the rest of the book.

Chapters 2 and 3 examine the predicament of the public: the people who are supposed to know for themselves, to exercise their reason, in the face of data-driven surveillance. Focusing on the Snowden affair, I argue that technologies of datafication often provoke paranoid and otherwise speculative forms of public knowledge and political participation. Ideal norms like transparent governments and informed, rational publics flounder when confronted by technological systems too large, too complex, and too opaque for human scrutiny. The Snowden files, and the electronic surveillance systems they describe, are thus recessive objects: things that promise to extend our knowledge but simultaneously manifest the contested and opaque nature of that knowledge. For both the American public and the intelligence agencies themselves, the surfeit of data provides not the clarity of predictive certainty but new pressures to judge and act in the face of uncertainty.

Chapter 4 then turns to self-surveillance and its promises of personal empowerment through the democratization of big data technologies. Paradoxically, this narrative of human empowerment is dependent on the privileging of machinic senses and automated analytics over individual experience, cognition, and affect. These new technologies for tracking and optimizing one's daily life redistribute the agency of knowing in ways that create new labors and dependencies. The chapter further traces how the Quantified Self is scaling up to the Quantified Us. Systems of fabrication first created for individual self-knowledge are gradually integrated into the wider data market, opening up new avenues of commercialization and control.

Chapters 5 digs into concrete techniques of fabrication, namely, how uncertainties surrounding terrorism and its attendant data—emails, web browsing, phone calls—are crafted into data-driven insights. Beneath and between the supermassive streams of data and metadata, improvised heuristics of speculation, simulation, and deferral help produce actionable knowledge claims. These are furtive sites in which specific forms of speculation and estimation are forged into sufficiently true evidence. Such techniques, so fragile and improvised when examined up close, ride the waves of broader technological fantasies around objectivity and progress. Chapter 6 shows how self-surveillance is presented as a historically inevitable step towards a vision of posthuman augmentation that I call "data-sense." The imperfections and contradictions of technological factmaking are thus consolidated into a normative demand that human subjects know themselves through data and thereby learn to live and reason in ways that help advance its self-fulfilling prophecy of better knowledge.

The book thus examines a number of different junctures in the social life of technologies of datafication. Chapters 4 and 5 offer a closer look at the technological side of how personal data is produced and leveraged, and their implications for human judgment in concrete sites of decision-making. Chapters 2 and 3 are more focused on how these technologies interact with existing political realities, challenging popular norms and expectations around rational publics and governments. Bookending these analyses are chapters 1 and 6, which attend to the underlying fantasies about technology and society that shape these specific practices of speculation.

Taken together, these scenes of datafication demonstrate the duality of fabrication and how the pursuit of data idealizes—and undermines—the figure of the good liberal subject. Edward Snowden justified his whistle-blowing of electronic surveillance programs with the argument that the American people must learn the truth about their own datafication. Yet how can the public fulfill its Enlightenment duty—*sapere aude!*—to have the courage to use one's own understanding—when systems of datafication recede "out there," beyond the horizon of individual experience and knowability? The Quantified Self community explicity cites history: the ancient Delphic maxim *gnothi seauton*, "to know thyself." But that knowing involves brokering a very different relationship between the self that knows, the knowledge that is allegedly their "own," and the machines that make it all possible. It is precisely such messy, speculative moments that matter for how standards of truth are being transformed. They are zones of transition, where new ways of proving and truth speaking are accorded the status of "sufficient" certainty to meet highly practical exigencies.[11] It is worth remembering that big data's "bigness" is not a matter of absolute thresholds but a relative one where qualities such as the volume and variety of the handled material exceed older bottlenecks and human limitations.[12] Yet those fleshly bottlenecks had served a function: they had slowed things down long enough for the exercise of judgment, debate, and accountability. Those opportunities for human intervention are now being systematically disrupted and overwritten. Algorithms, especially because so many tend to be classified or proprietary, themselves become sources of uncertainty because they introduce layers of mediation that become opaque to human scrutiny.[13] Across state and self-surveillance, the pursuit of better knowledge constantly reframes the distribution of rights and responsibilities across the subject meant to know, the ever-growing panoply of machines surrounding that subject, and the commercial and governmental interests behind those machines.

Technological Defaults

The stakes of data-driven fabrications, of the changing standard of what counts as truth, cannot be confined to epistemology, but relate directly to questions of power and justice. This is a truism that bears repeating, for postwar technoscience as industry and vocation has accumulated an enduring myth of depoliticization. The idea that one merely pursues objective truth, or just builds things that work, serves as a refuge from the messiness of social problems.[14] The question of what counts as knowledge leads directly to questions of what counts as intent, as prosecutable behavior, as evidence to surveil and to incarcerate? What kind of testimony is made to count over my own words and memories and experiences, to the point where my own smart machine might contest my alibi in a court of law? What constellation of smart machines, Silicon Valley developers, third-party advertisers, and other actors should determine the metrics that exhort the subject to be fitter, happier, and more productive?

Big data and smart machines push the bar toward a society in which individual human life, sensory experience, and the exercise of reason is increasingly considered unreliable. At the same time, what might otherwise look like flaky numbers, prejudiced estimates, or dubious correlations are upgraded into the status of data-driven insights, black-boxed from public scrutiny and fast-tracked to deployment. This book argues that fantasies of machinic objectivity and pure data work to establish datafication as a technological default, where ubiquitous surveillance at both populational and personal scales are presented to the public as not only a new and attractive technology but also an inevitable future.

This default is a bottleneck for the imagination, for the ability to devise and build consensus around the kinds of policies, ethical imperatives, social norms, and even technologies that might help us manage the consequences of datafication. The answer is not, however, to run in the other direction, to romanticize humanity before the internet. Such atavism reproduces the myth of a sovereign, independent subject, one which might be resuscitated simply by detoxing ourselves from technological influence. There is no returning to the mythical time of the good liberal subject, and the transparent disclosure of the ever-expanding webs of datafication will not in itself restore the capacity for rational self-determination. Instead, we should ask of data's promise of better knowledge: What good does it really do to "know"? What other conditions, beyond the often narrowly defined metrics of accuracy and efficiency, are necessary to ensure that knowledge empowers the exercise of human reason? How can those conditions be protected as the process of knowing is increasingly overtaken by opaque systems of datafication? As I elaborate in the conclusion, asking these questions requires disrespecting the stories that data tells about itself, to refuse its rationalization of what looks like objectivity or progress, and to hold technology accountable to standards that are external to its own conditions of optimization. To refuse technology's rules of the game is to refuse the steady entrenchment of a rationality where datafication and its knowledge claims are increasingly neither by or for "us"—the human subject, the individual, the rational public—but pursues its own economic and technical priorities.

Ultimately, this book is a story of how datafication turns bodies into facts—a process that aspires to a pure and pristine objectivity but, in practice, creates its own gaps and asymmetries. The ambitious projects for state and self-surveillance reveal crucial gaps between the narrow reaches of human knowability and the vast amounts of data harvested by machines, between the public that is supposed to know and the institutions and machines that are meant to know it in their stead, between the practical capabilities of data technologies and the wider fantasies that give them legitimacy. In each one, we find troubling asymmetries in how different bodies are treated to different kinds of factmaking. If data expands the vistas of human action and judgment, it also obscures them, leaving human subjects to work ever harder to remain legible and legitimate to the machines whose judgment they cannot understand. Caught in an expanding and consolidating data market, we cannot simply seek more and better knowledge but must rethink the basic virtues and assumptions embedded in that very word. What kind of good does knowing do? Or, rather, what must our knowledge look like that it may do good? And who are we, with what kinds of capabilities and responsibilities, with what role to play in a data-driven society? As the truth of who we are and what is good for us is increasingly taken outside ourselves and human experience, the figure of the human subject—which, Foucault had warned, is a young and temporary thing[15]—is flickering uncertainly, unsure of the agency and moral responsibility we had worked so hard to attach to it.

1

Honeymoon Objectivity

In 2014, a baby-faced, twenty-two-year-old entrepreneur named James Proud crowdfunded a sleep-tracking device that promised to automatically monitor sleep patterns, provide a numerical score, and make recommendations for sleep behavior. That such functions were already available did not escape Proud. Beddit, a sleep sensor that we will revisit in chapter 4, had been crowdfunded a year before and already released to its backers. In response, Proud chose to emphasize his device's "simple, uncomplicated and useful" qualities; designed as a slick, minimalist off-white orb, it would merge invisibly into the everyday flow of attention and reflection. "We believe technology needs to disappear," said Proud; "everything in [our device] is just designed to fade away."[1] It would carry an equally simple and no-brainer name: Sense.

In 2017, James Proud, now twenty-five, announced the end of Sense.[2] Panned by some tech reviewers as a "fundamentally useless" object[3] and a glorified alarm clock, the device never quite delivered the quiet transformation of everyday life that its creator aspired to. Fundamentally, it proved not very good at making sense of human sleep. Users reported that any deviation from the presumed sleep scenario—for instance, a pet snuggling up in bed—would throw the device off entirely. The chaos of everyday life rarely conformed to the expectations of the tracking machine, even as its selling point was that it would discover truths about us that we cannot perceive ourselves. As Proud's team wound down operations, users began to report that their Senses were losing functionality. The orbs went mute and deaf to the data around them, a small monument to the unfulfilled promises of new technologies.

Technologies of datafication reconfigure what counts as truth and who—or what—has the right to produce it, and not simply through the success of indisputably superior machines or even their mundane and ordinary operations in concrete practice. As relatively nascent technologies fast-tracked to the status of a global buzzword, the very idea of big data and smart machines—the spectacular keynotes and product launches, the anticipatory rhetoric, the science fiction, the projected future functions—operates as a social actor in its own right.

Proud's story, after all, is a common one. The popularization of self-surveillance technologies followed decades of anticipation (and disappointment) about a future that was always advertised as just about to arrive—a "proximate future."[4] A computing magazine included "smart appliances" in a 2007 piece about the "biggest technology flops," deriding the "bubble" around smart appliances back at the turn of the century. "The bubble burst, and we haven't heard much about intelligent appliances since," the article said.[5] That very year, the Quantified Self (QS) community would emerge in Silicon Valley; by 2011, the Internet of Things was back on the forefront of the imagined future, featuring in the tech advisory firm Gartner's influential "Hype Cycle for Emerging Technologies" report for the year.[6] Yet this return to the spotlight of the imminent future was not necessarily built on clear and proven cases of better knowledge. The rapid growth of the self-surveillance industry provoked public skepticism, academic research alleging negligible or backfiring effects,[7] and even lawsuits challenging the basic accuracy of popular measuring devices (namely, *Brian H. Robb v. Fitbit Inc., et al.* 2016). The broader industry of smart machines was no better off; one internet-connected juice maker cost $400, but its proprietary juice packs turned out to be just as squeezable by hand. A smart lock automatically updated over wireless connections and then locked users out of their homes; smart salt shakers promised voice-activated controls but were unable to grind salt. The proximate future was cobbled out of Eric Kluitenberg has called imaginary media:[8] prototypes depicting impossible realities, products sold on the basis of never-quite-actualized functions, artists' sketches, and bullish press conferences. Even as they malfunction and disappoint, they help drag impossible functions and nonexistent relations into the realm of the sayable and thinkable. Consumers are asked to buy into the latest gadget in anticipation of its future ubiquity, to install software for its future functions, and to celebrate prototypes for their glimpse of what, surely, must be just around the corner.

Technologies of datafication seize the authority to speak truth not by achieving the improbable hype gathered around it but by leveraging those lofty goals to mobilize the public, siphon funding, and converge collective imagination. Technology thus operates not merely as tools and functions but also as a panoply of fantasies—about machines that know us better than we know ourselves, about predicting the future with pinpoint accuracy, and about posthuman cyborgs and Big Brothers. To say "fantasy" does nothing to undermine the unique importance of material facts (as if fantasies could be sustained, or even generated in the first place, without the affordances of concrete things!). But it does mean tracing the ways in which data-driven surveillance seized its claim to knowledge by mobilizing projections and estimations about technology and the future world that will necessitate those technologies. While tracking devices such as Proud's were crafting an optimistic technofuture animated by consumerism, tugging on the broader imaginary of posthuman augmentation, state surveillance systems were warning of a future that must not happen, predictions of crime and terror that must be snuffed out through strategies of "zero tolerance." Across both cases, fantasy takes half a step outside present reality not to escape from it but to all the more effectively guide it.[9] Žižek once observed of the dystopian science-fiction film *Children of Men*[10] that

> Hegel in his aesthetics says that a good portrayal looks more like the person who is portrayed than the person itself. A good portrayal is more you than you are yourself. And I think this is what the film does with our reality.

The market projections, promotional rhetoric, bullish claims, and dystopian warnings surrounding datafication today are precisely the little doses of fiction used to make sense of these technologies and the knowledge they promise. Such beliefs are not reducible to "intellectual mistakes" by naïve or ignorant subjects. This (mis)recognition of what technology does and could do, the benefit of the doubt and the doubtful benefits, is so often a crucial part of getting technoscience off the ground.[11]

When New Technofantasies Are Old

Round about stood such as inspired terror, shouting: Here
comes the New, it's all new, salute the new, be new like us!
And those who heard, heard nothing but their shouts, but
those who saw, saw such as were not shouting.
—Bertolt Brecht, *Parade of the Old New*

Fantasy, in this collective, commercialized, politicized form, is never
truly free-form. Datafication often falls lockstep with familiar narratives around machines and rationality, tapping into that modern drive
to order the world as a taxonomy of facts for a sense of legitimacy and
plausibility. As Lauren Berlant shows, these familiar anchors help stitch
together the contradictions and disappointments of technology, the gaps
between knowledge and uncertainty, into a sense that "the world 'add[s]
up to something,'" even when that belief is constantly displaced and
disappointed.[12] The paradox throughout this book is that technologies
of datafication rely so heavily on the imagined legacy of the Enlightenment, and its particular alliance of objectivity, human reason, and
technological progress, even as its deployment threatens to destabilize
the presumed link between information, human Reason and democratic
freedoms. Since its emergence over the mid-nineteenth century, the
thoroughly modern concept of technology has depicted a world whose
every aspect stands ready to be flattened, standardized, and turned into
problems that the ceaseless march of new inventions would render into
objectively optimal states.[13] The fabrications explored in this book leverage what we might call honeymoon objectivity: the recurring hope that
with *this* generation of technological marvels, we shall establish a universal grounding for our knowledge, a bedrock of certainty, a genuine
route to the raw objective layer of the world around us. By invoking this
long quest, tracking technologies are able to draw together their own
imperfections, uncooperative material conditions, incompatible and
otherwise resistant humans into a seductive vision of better knowledge.

The objectivity invoked by data-driven surveillance constitutes no
rigid dogma but a sedimented range of attitudes and affects embracing
a distinctly modern way of thinking and feeling about knowledge. As
extensively chronicled by Lorraine Daston and Peter Galison,[14] older

renditions, such as the scholastic *obiectivus/obiectiv*, generally involved definitions starkly different from the modern. Even in Kant, objective validity meant general "forms of sensibility" that prepare experience, while the "subjective" referred to specific and concretely empirical sensations. It is only during the nineteenth century that the now-familiar juxtaposition emerges: a neutral, "aperspectival" objectivity as the privileged instrument toward truth and scientific inquiry and biased, unreliable subjectivity as its nemesis.[15] By the later nineteenth century, Daston and Galison identify a dominance of "mechanical objectivity": a regulative ideal that called for the elimination of the human observer from the process of data visualization. The critical impulse for these conceptualizations was, of course, the advent of photographic technology, which provoked new theories and standards for what counts as visual truth and who (or what) might be best equipped to produce it. Photography – despite its own long history of manipulation and contested meanings – thus spurred new linkages between automation and objectivity, producing the ideal where "machines [would be] paragons of certain human virtues" precisely by ridding themselves of human subjectivity.[16]

The public presentation of data-driven surveillance leverages these older ideals of objectivity, and the cultural capital it had accumulated through traditions of scientific inquiry. In self-surveillance's effort to map the microbiome or record every moment of sleep, we find a conception of the body as an aggregation of correlations. Health, productivity, and happiness are broken down into a set of hidden but logical relationships that machines might read and catalogue—the same kind of correlations that might help predict the lone wolf terrorist, enabling an orderly distribution of risk and suspicion across the population. In this cultural imaginary, the world is an indefinite archive, and the machines of tomorrow, if not of today, will be up to the task of cataloguing it.

All this is not to say that the Enlightenment ever bequeathed a singular doctrine about technology and reason or that different practices of datafication share a totally coherent conception of a value such as objectivity. As Lorraine Daston herself noted, each historical rendition of objectivity expresses not some immutable quality rooted in natural law but a mélange of aspirational values that happen to occupy (or, at least, contest) a normative position at the time.[17] Indeed, honeymoon objectivity describes that recurring tendency to *claim to* make new advances toward such an immutable quality, even as the kinds of data actually produced might diverge significantly from this vision. Technologies of datafication do not subscribe neatly to any single definition but cobble together different popular imaginations of what objectivity looks and feels like. The central proposition of mechanical objectivity—the preference of nonhuman processes over subjective ones for the reliable production of knowledge—is retained as a basic article of faith, but one that is routinely transgressed and compromised in practice. The messy and flexible ways in which the virtue of objectivity is "localized" onto self-surveillance cultures reflects, above all, how broad and pliable the word has become. Like culture, objectivity exhibits a certain "strategic ambiguity."[18] Its many possible permutations allow a wide variety of interpretations and attitudes to rally behind a common banner, where more fine-grained definitions might have splintered them. Thus, the fantasy endures to pass on its allure to another institution, another machine.

Pure Data

If the pursuit of objectivity, in all its strategic ambiguity, is the well-advertised benchmark of data-driven surveillance, an equally crucial question is: What kind of regime of knowledge, what kind of social order, is it meant to deliver? This book argues that the many articulations of data's benefits, capacities, its factmaking powers, revolve around a mythologization of data as pure and purifying. This pattern emerges not so much in efforts toward the technical definition of data, but in the public discourse, where the very question of what data *is*—or, rather, what can data *do*—again involves a messy plurality of ideas and dispositions. Data, fact, information, and knowledge are often conflated such that they are either seen to naturally follow on from each other, bolstering a sense of legitimacy.[19]

Three years after the first Snowden leaks, BBC4 released a documentary titled *The Joy of Data*. Its host, the mathematician Hannah Fry, boiled it down to a pyramid. From bottom to top, she explained, data is "anything that when analysed becomes information, which in turn is the raw material for knowledge—the only true path to wisdom."[20] Fry left unsaid what exactly knowledge and wisdom were, but the hierarchical relation was clear: the raw objective facts gathered through new technologies would serve as the foundation for better knowledge. This DIKW pyramid—data, information, knowledge, wisdom, in ascending order—is a fixture in many computer science textbooks. Underpinning it is a world in which everything we can or need know is reducible to positivist facts, and by descending to this atomic layer, we will be able to recover objective data for any problem.[21]

In this articulation, data and knowledge are inseparable bedfellows. Data is the ubiquitous ingredient in the buoyant dreams of better knowledge, the object unto which the hopes and fears of technological and epistemic possibility are invested. In its most elementary form, it is described as raw data: data generated by the machine but yet to undergo "secondary" processes of statistical analysis, cleaning, visualization, aggregation, and so on. It is data fresh out of the sensor, with no artificial additives. In this telling, raw data is seemingly anterior to analysis, classification, and attribution of meaning. The valorization of raw data is intimately connected to self-surveillance's vision of empowerment through objective knowledge. In 2015, one QSer suggested raw data access as one of the three "freedoms of personal data rights":[22]

> Without raw data, we are captive to the "interface" to data that a data holder provides. Raw data is the "source code" underlying this experience. Access to raw data is fundamental to giving us the freedom to use our data in other ways.

Similar sentiments were expressed by a host of prominent commentators, including QS co-founder Gary Wolf.[23] The widespread implication of raw data's nonmediated nature translates into the fantasy of data as a purifying agent: a technology that will produce knowledge stripped clean of politics, of human bias, and of troublesome differences in opinion and establish the clear and rational path forward. Yet, as numerous scholars have pointed out, the very idea of raw data is an oxymoron.[24] Data only becomes data through the human design of each algorithm, relational database, and deep learning system—although there are important differences in how much detail is determined by manual design and judgment and how much is left up to machine learning.[25] Data is no thing-in-itself that exists prior to observation but something to be "achieved"[26] through a concerted process of production that can never rid itself of human subjectivity and sensibility.

The dangerous consequence of this aspiration to purity is that the human, social, historical, and moral aspects of data's fabrications are invisibilized—allowing familiar kinds of speculation and prejudice to reenter by the back door. Consider the effort to predict and intercept terrorists before they can cause harm. Chapter 5 examines known cases of sting operations where certain individuals—predominantly young, Muslim, of Arab descent, male—are marked out by state surveillance apparatuses for fabrication. Driven by a moral and political injunction to "zero tolerance," in which even a single terrorist attack is an unacceptable failure of prediction and control in the wake of the September 11 attacks, counterterrorist operations do not simply wait for the data but actively work to produce the necessary proof. Thus, in the case of Sami Osmakac, FBI undercover agents supply the individual with money and the weapons and explosives to be purchased with that money, and coach him each step of the way until arrest can be justified. Meanwhile, the Snowden files reveal the surveillance programs themselves to be inevitably human. Analysts from the National Security Agency (NSA) speak of "analysis paralysis" and the struggle to handle supermassive volumes of data, while placeholder entities, such as "Mohamed Badguy" and "Mohammed Raghead," for database-search interfaces reflect the all-too-human, all-too-crude underside of sophisticated data-driven systems. Criticisms of datafication have often invoked labels such as data doubles and doppelgängers to warn against how individual self-expression is being replaced by alternative identities recomposed from data extraction.[27] Alongside such "copies," we also find a variegated ecosystem of speculative entities: the Osmakac that might have been, the Raghead in the database. Here, datafication provides no mathematical certainty but a range of possible outcomes and correlations to legitimize highly anticipatory forms of surveillance, judgment, and incarceration. The desire for epistemic purity, of knowledge stripped of uncertainty and human guesswork, ends up with concrete practices that draw perilously close to the imaginations of purifying the nation and the body politic. Suspected terrorists, brown or white, religious fanatic or ethnonationalist, end up subject to very different forms of datafication even as the technology promises a neutral illumination of truth.

The idea that raw data can access an orderly and calculable reality stripped of historical forces, social constructions, and political disputes translates into a converse fantasy: that the individual body will be purified of the elements that impair its health and productivity, and the body politic too will be secured as a transparent whole. Yet a central argument of this book is that such a fantasy of epistemic purity—of knowledge untainted, of complete certainty—itself carries political and moral biases. The belief in raw and untainted data begets not only an excessive reliance on algorithmic factmaking but also extends the older and deeper cultural desire for sorting the world into stable and discrete pieces. The recurring temptation: What if we could predict and eliminate every bad apple, every violent individual, and every criminal intent? What if we could maximize everything good about our bodies, expel all the toxins, cut off the bad friendships, and optimize every habit? Just as the pursuit of better knowledge through datafication entails a social shift in what counts as objectively true, the collective faith in the purity of data entails using the data to try to bypass important political and moral questions, to try to purify bodies through technological solutions.

The Groundless Ground

The mythologization of pure data puts into ironic relief the original Latin: *data*, meaning "that which is given." Today, (raw, big) data's privileged position in objective inquiry and knowledge production seeks to normalize into the woodwork, becoming "something we would *not* want to deconstruct."[28] It has been called the "cathedral of computation," or a faith in "computationalism": the fantasy that data simply *is* and shall provide a reassuring grounding for everything else that trouble us.[29] This faith has immediate and practical rewards. If datafication promises objectivity, impartiality, and predictivity, all these epistemic characteristics add up to a valuable sense of *stability*. On one hand, specific processes of data-driven analytics work within narrowly defined parameters where inputs may be standardized, modeled, and otherwise manipulated. In other words, the algorithm's truth claim itself relies on a set of grounding assumptions about the world out there and its methodological relation to data—assumptions that it agrees not to question to get the job done. At the level of data as a broader, popular imaginary, telling here is the enduring popularity of a rather naïve extrapolation of Shannon's law of information: the idea that we can progressively eliminate uncertainty in all situations through the addition of information (which are themselves certain), each of which would reduce uncertainty in varying amounts. Both as a technical procedure and as a social imaginary, datafication thus consists not simply of truth claims but also the normalization of a new kind of grounding for knowledge claims.

This grounding, this social basis of felt certainty, was precisely the subject of Wittgenstein's final, incomplete work. In it, he asks, "What is entailed in the simple phrase, 'I know'?" There is a curious masking function: the act of saying "I know this is a tree," for instance, does not establish any comprehensive or objectively certain proof that I really do know. Yet we trust such claims on a regular basis, tacitly agreeing not to question them too far; after all, only philosophers bother to hold regular debates revisiting whether trees really exist. Our knowledge claims provide no indisputable foundation. The very act of saying "I know" seeks to "exempt certain propositions from doubt," to agree to not to look too closely.[30] This infrastructure of common sense is what Wittgenstein provisionally labeled world-picture, *Weltbild*: models that allow us to cope with the world, to make certainty and judgment possible.[31]

Yet herein lies an unresolvable paradox at the heart of claims to better knowledge: the groundlessness of the ground itself, or, the ways in which the demarcation of what "counts" as good knowledge is ultimately arbitrary. Wittgenstein comments that "at the foundation of well-founded belief lies belief that is not founded"[32]—precisely because to claim "I know" is an act that removes its contents from the game of proof and justification. This arbitrariness is well exposed by young children not yet versed in the unspoken boundaries of the language game: "But how do you know it is a tree?" "Well, it has branches, a trunk, some leaves." "But how do you know those are branches?" "Well, if you look at an encyclopedia—" "But how do *they* know those are branches?" and so on until the frustrated adult snaps: "We just know, okay?" We might reasonably dispute whether such grounding is truly ground*less* or simply deferred and bracketed in sufficiently complex ways that it can be presumed in ordinary contexts. For our purposes, the two options have the same consequence. For ordinary subjects, navigating their everyday life, pressed to judge and form opinions about things increasingly be-

yond their phenomenological horizon (such as a vast and secret government surveillance system or bodily physiological processes beyond the
human sensorium), there is a practical need to make or accept knowledge claims, to not question their indefinite regress as the child does.
Hence, Wittgenstein inverts the typical model of proof to say that when
I say I know this is a tree, "it is anchored in all my *questions and answers,*
so anchored that I cannot touch it."[33] To question that it is a tree, or to
question how I can know any such thing, shakes too much of the edifice
built above it that it typically becomes *unreasonable* to question it.

It is this ground that is being reconfigured when counterterrorism efforts blur lines with sting operations or when self-surveillance technologies are promoted as superseding human memory, cognition, and other
"natural" means of datafying the natural world and their own body.[34]
The following chapters examine how specific and often-imperfect techniques for prediction and analysis become valorized as objectively superior knowledge. Meanwhile, a growing set of assumptions—about the
nature of data, the value of human thought and machinic calculation,
the knowability of the world out there and the human body as an information machine—become "set apart" and invisibilized, melding into
the background of everyday experience and of public discourse on data-
driven knowledge. Across both state and self-surveillance, the material
objects of datafication constantly seek to sink into the background of
lived experience—mirroring the disappearance of data as a social construction deep into the ground. The NSA's data collection occurs not at
the embodied sites of personal communications but through undersea
fiberoptic cables, restricted-access data centers deep in the Utah desert,
or buildings hidden in plain sight as a brutalist New York skyscraper.[35]
Self-surveillance devices, at first thrust into the spotlight as delightful
novelties, are increasingly seeking to recede into the realm of habit and
unnoticed ubiquity—where their influence on individuals no longer
needs to be justified through active and spectacular use. Datafication, in
short, seeks to become our groundless ground.

The groundless ground constantly encourages those who live on it
to forget how contingent it is. Pointing to the most basic elements in
scientific and mathematical reasoning, Ian Hacking speaks of "styles of
reasoning": nothing even so complicated as a system of measurement
or a law but something as elementary as, say, the "ordering of variety

by comparison and taxonomy."[36] Like the epistemic qualities Foucault charted in *The Order of Things*, these basic tendencies rarely come up for debate even as theories and ideologies are toppled. They change far more slowly and so provide stable grounding that allow us to perceive a fact as fact in the first place.[37] In the data-driven society, such styles of reasoning govern how we relate a number (an algorithmically generated expression of reality) to the body's sensory experience, to conscious human testimony, and to other sources of truth. It governs how bodies are turned into facts: what kinds of bodies become eligible for what kind of datafication and how different bodies are treated to different kinds of factmaking processes. To identify the groundless ground as ultimately arbitrary and conventional is not to say that they are therefore illegitimate; such fabrication is, once again, a normal part of the social existence of things.[38] What it does mean is that data's claim to better knowledge is not a given, and neither are the forms of factmaking they bequeath on society. There are important political and moral choices to be made around what kinds of authorities should serve as the groundless ground and what kinds of data, machines, and predictions should count as looking and feeling like truth.

The Data Market

The epistemic fantasies of datafication matter—not when or if they deliver on all their promises but in the present, where the mobilization of collective belief in those fantasies transform what counts as truth and certainty. The patterns and tendencies specific to contemporary state and self-surveillance stem from two important tendencies in big data analytics: indifference and recombinability. Big data analytics are predicated on the ceaseless production of data indifferent to its specific nature and without a rigid presumption of its utility—because this data will always remain open further exploratory analyses, recombining different datasets and analytical methods to discover unforeseen correlations.[39] This is indifference to causality in favor of correlation; indifference to "intelligence," in the sense that the data is collected without the prior establishment of an interpretive context; and, as subsequent chapters show, indifference to the human experience of the world and that context of everyday living. To be sure, indifference does not mean neutrality.

Even as many aspects of the analytical process become automated and left up to learning machines, the design of those learning processes and the initial identification of the kind of data to be gathered renders it an "interested," if not deliberately biased, process.[40]

One such driving interest is precisely the manufacture of usable, justifiable certainty. Algorithms, as Louise Amoore puts it, "allo[w] the indeterminacies of data to become a means of learning and making decisions."[41] Messy data, extracted from lived experience and social reality and reordered into machine-readable form and modeled into a comprehensible pattern, are leveraged to produce truth claims that are not simply true or false but are carefully packaged expressions of probability that harbor uncertainty by definition. These are deployed and sold as freely transportable systems for generating "insights" across different social problems. To begin with, technologies and products are often crafted for fairly specific purposes. But that very act of measuring often involves recombining whatever data that can be conveniently acquired until a useful correlation (i.e., a profitable payload) is discovered, and it is also common that such data collection later leads to new and formerly unimagined kinds of predictions. Thus, the sex-tracking app Spreadsheets measures "thrusts per minute," a largely pointless value for any human assessment of sexual intercourse but one that the movement sensors on a typical smartphone are well equipped to provide. Such sensors, originally implemented for distinct features (such as the use of accelerometer and gyroscope data to allow portrait/landscape orientations on smartphone screens), create new affordances for the business of tracking. Big data analytics often has "no clearly defined endpoints or values,"[42] precisely because its profitability hinges on the expectation that any given algorithm, any process of datafication, might potentially be exported as a standard procedure for an indefinite range of activities (and thus business opportunities).

State and self-surveillance, despite their many local differences, thus participate in a wider, cross-contextual data market. The seemingly technical tendencies of indifference and recombination work to encourage a particular set of political and economic realities. The optimism that any and every process can be improved through datafication constitutes a voracious impulse that reveals big data's fundamental affinity with capitalism's search for continual growth.[43] The larger the userbase, the more data to be extracted, which not only refines the primary analytics but also increases the possibility of recombining that data for new uses (or for selling them on to third-party buyers). Thus PredPol, the prominent predictive analytics system for law enforcement, borrows from existing earthquake modeling techniques.[44] Palantir, a private data analytics company, was born out of funding from In-Q-Tel, the venture capital arm of the CIA, and then sold its products back to intelligence agencies.[45] It has subsequently begun to reach out to corporate clients, such as American Express and JPMorgan Chase, demonstrating the ease with which antiterrorist technologies and antiterrorist funding can be leveraged for civilian surveillance.[46] Fitbit, one of the most popular tracking devices during the mid-2010s, is piloting partnerships with insurance companies,[47] and a significant minority of products have been reported to share data with third parties,[48] following exactly in the footsteps of social media platforms' journey to profitability.

The data market advances what has been called "surveillance capitalism": the work of making the world more compatible with data extraction for recombinant value generation.[49] This perspective situates what is promoted as a technological breakthrough in a longer historical cycle of capitalist "logics of accumulation,"[50] including the postwar military-industrial complex.[51] In effect, the data market constitutes an early twenty-first-century answer to capitalism's search for new sources of surplus value. Here, new technological solutions are presented as (1) a universal optimizer, which is hoped to short-circuit existing relations of production and maximize the ratio at which labor power is converted into surplus value, and (2) itself a commodity, which may be hyped up for a new round of consumerist excitement.[52] Surplus value is located not so much in the optimization of prices and goods sales but in the optimization of data extraction and refinement.[53] The "profit" at the end of this process is sometimes obviously commercial, as in targeted advertising and the direct selling of consumer goods. But the profits or uses of surveillance capitalism must also be counted in the biopolitical sense, wherein state securitization seeks to identify and manage the normal population or the individual consumer is enjoined to render themselves more attractive to algorithmic decision-making systems through techniques of self-optimization. The constant traffic and recombination of data thus entail an ever-wider range of situations in which data may substitute or override the claims of physical bodies, conscious subjects, and lived experience.

In commercialized spheres, such as self-surveillance (and even in state surveillance, where the drive to datafy produces opportunities for lucrative government contracts for private firms), the logic of accumulation is the engine that animates datafication's promise of better knowledge. In this light, the ongoing demotion of human knowing in favor of machinic measurement and data-driven insight is not simply an intellectual argument but a variation of what David Harvey called "accumulation by dispossession": the seizure of assets to release at extremely low costs, producing new opportunities for profit that predictably benefit those with incumbent capital.[54] The more devalued human intelligence, the better for selling artificial intelligence. With datafication, the deep somatic internality of the self—my desires, my intentions, my beliefs—are opened up for revaluation on terms distinctly favorable to new products and systems of datafication. Exhorting the virtues of self-surveillance requires downgrading the reliability of human memory and cognition, such that the smart machines—and the new industries of hardware sales as well as the subsequent recombination of that data—is seen as necessary to true self-knowledge.

These trends extend long-standing tendencies in the history of surveillance, both digital and otherwise. After all, Foucauldian discipline was never about the sovereign execution of coercive power through surveillance; it was itself a highly distributed and participatory practice pegged to the promises of knowledge and productivity. To be sure, embedded in the very word *surveillance*—composed from the French *sur* (above) and *veiller* (to watch)—is a specific relation: domination from "above" through optics. But alongside that straightforward image of Big Brother is a history of surveillance as a technique for producing truth, affixing subjects to the identities and roles prescribed by that truth, and, ultimately, *disciplining* subjects into general dispositions and ways of seeing. To ponder whether we are "panoptic" or "synoptic" or "post-panoptic" is to miss the broader continuity of that liberal principle in which subjects participate in their own surveillance through the internalization of a certain way of seeing.[55] The lesson shared across the panopticon, the ominous screens of *1984*, the highly visible CCTV installations in London's streets, is that what really matters is not (only) the active relation of a watching subject and the watched one but the generalization of the condition of being *under* surveillance—a condition that corrals the human body and all it does into a standing-reserve of evidentiary material for interpretation, recombination, and classification.[56] From this vantage point, what is fundamental to surveillance is not the image of an Orwellian coercive control but a set of processes by which my truth becomes defined by those other than myself through a systematic and standardizing mode of organization. Surveillance, in this sense, is inseparable from the history of large-scale communication technologies and often develops in lockstep with the reach of the latter.

This book asks what kinds of politics, what kind of subjectivity, becomes afforded through the normalization of these technological fantasies around objectivity and purity and through the cross-contextual expansion of the data market. In the data-driven society, "what counts as knowledge" so often ends up a question of what counts as *my* body, *my* truth, *my* eligibility for social services, *my* chances of being targeted for surveillance, *my* chances at a job . . . Even as the idea of big data bloomed into a ubiquitous buzzword, its ambiguous consequences continued to break out in accidents and scandals. Some were told through the popular annals of outrageous stories: the man who was fired by algorithms,[57] the African Americans categorized as gorillas by Google Images.[58] Other controversies were more wide-ranging and enduring, such as the Snowden affair itself. It has been described as the "data wars": the growing social conflict over how people's algorithmic identities are determined and by whom.[59] Like the culture wars, what is at stake is the distribution of labels and associations by which we can identify, sort, and make judgments on individuals.

The trouble is that even as big data and smart machines invoke the thoroughly modern and Enlightenment imagery of technological progress and societal reform, this generalization of indifferent and recombinant factmaking often serves to retrench politics and economics as usual. The mix of naïve liberal individualism and technocracy that fuels the visions of machine-optimized futures provides no fresh political vision for the distribution of resources or the organization of collectives. There is only the conceit that with new technologies, we can finally achieve a fully automated luxury capitalism. Indeed, the very idea of "optimizing" reflects one of capitalism's essential assumptions: that there is always another world beyond this one to plunder, that there is no end to expansion, and that we shall not run out of resources, of new conquests, new sources of value.[60] That capitalism, just like technology, just needs the next upgrade, the next invention, to really fulfill its pure vision of totally frictionless transactions and truly melt all that is solid into air. Dressed in the shining garb of technological novelty, datafication proves most of all the difficulty of proposing a coherent alternative to capitalism and the good liberal subject.[61] The push for datafication thus extends and depends on enduring fantasies around liberal values, even as its implementation often reprises old roadblocks and compromises. We now turn to one such impasse in the Snowden affair, where an unanswered question looms above all the debates around transparency and secrecy, surveillance and privacy: Can the public truly know for itself in the age of nonhuman technologies? If not, what kind of politics remains?

2

The Indefinite Archive

I know I am being watched; Edward Snowden told me so—although I cannot experience it for myself. This strange disjuncture spells out the problem: What does it mean to "know" technological systems that grow ever larger and more complex and yet are concealed from the human subject?

In April 2015, Snowden—having sought asylum in Russia—agreed to an interview with the comedian–cum–talk show host John Oliver.[1] Oliver had brought a dose of realism for the young idealist: Do you think the American people now possess the knowledge you have given them? Do they even know who you are? His clip showed a series of passersby at Times Square: "I've never heard of Edward Snowden," said one. "Well, he's, um, he sold some information to people," ventured another. The knowledge that Snowden had risked his life to impart seemed to have dispersed into the crowded streets—visible here and there but in piecemeal and confused forms. Oliver offered consolation in textbook deadpan: "On the plus side, you might be able to go home, 'cos it seems like no one knows who the fuck you are or what the fuck you do."

* * *

Timothy Morton writes that the Anthropocene presents humans with a proliferation of *hyperobjects*: things with such broad temporal and spatial reach that they exceed the phenomenological horizon of human subjects.[2] Images of endless (but equally fast-disappearing) ice sheets, floating garbage islands in the ocean, or statistical projections of planetary destruction, each evokes an uncanny sense of displacement: phenomena that seem to defy human scales of interpretation and yet demand that we reckon with them here and now. A variation of this question is posed by the Snowden affair. How can we "know about" technologies of datafication—the "we" being the amorphous yet enduring ideal of the public? Through Snowden's leaks, the public is called on to know for itself, a duty that it has borne since the Enlightenment. Its slogan, *Sapere aude*, bestowed by none other than Immanuel Kant, calls for individuals to have the courage to use their own understanding—to think and know for themselves.[3]

Notably, Kant ended his text with the comment that such knowing achieves nothing less than "man, who is now *more than a machine*."[4] Although he meant something a little different by that phrase, these words resonate with the tension between the good liberal subject and the datafied body. The discourse of big data presents the two as complementary. Yet the Snowden affair also raises the problem of human knowability in the age of data overproduction. Even as Snowden delivers essential information to the public, the technological systems in question increasingly defy human comprehension, producing an unstable gap between what the public is expected to know to function as rational subjects and the limits of their phenomenological horizon. Genuine and rare as they are, Snowden's documents also needed to be fabricated into the status of public knowledge. I argue that this process exposes underlying contradictions between technologies of datafication and the liberal ideal of open and transparent information.

In the Snowden affair, these problems are expressed through sets of common binaries—secrecy and transparency, knowledge and ignorance—which are then regularly transgressed, diluted, and short-circuited. This ambivalence is embodied by the Snowden files: the voluminous cache of secret documents whose leakage sparked the affair. They serve as evidence but also as objects of mystery. They are credited with radical transparency but also generate speculation and uncertainty. They establish their status as irrefutable evidence by appealing to the aesthetics of quantification but also normalize a certain kind of paranoia. The files constitute what I describe in the next chapter as "recessive objects": things that promise to extend our knowability but thereby publicize the very uncertainty that threatens those claims to knowledge. Recessive objects materialize the precarious and arbitrary nature of the groundless ground, showing how the very effort to mobilize technology for truth requires putting uncertainties to work.

To trace the public life of the Snowden files is to examine the ways in which the public is called on to "know about" hyperobjective technological systems. This chapter focuses on the Snowden files and the problem of knowing about state surveillance technologies. It forms a duet with chapter 3, which considers how the public and the state seek to "know through" these technologies the dangerous world of twenty-first-century terrorism. Together, they pose the question: How can the public know for itself the vast, expansive world out there—the world both of terrorism as unknown dangers and of surveillance itself as a pervasive technological system?

Data at Large

20 July, 2013. Journalists at *The Guardian* descended into their company basement, power drill and angle grinder in hand. Observed by two British state officials, they duly carried out the task at hand: the physical destruction of a laptop computer. The Apple MacBook Pro had contained top-secret files about American and British state surveillance activities, leaked to the left-leaning paper by Edward Snowden. Although the material had already been studied and reported on globally, the state insisted on this act of symbolic dismemberment.[5]

It is fair to assume that everybody present understood how parochial a ritual they were performing. As the laptop expired under a cloud of dust and debris, the Snowden files had already circulated to a global network of journalists and activists, including *The Guardian*'s own offices in the United States.[6] Distributed through mundane USB drives to a smattering of journalists a month prior, some of the files had already become the biggest news stories of the year.[7] Still, the ceremony was correct about one thing: the overriding importance of the files as material evidence. They detailed activities such as the bulk collection of telephone and email metadata from domestic populations at a massive scale and made available as searchable databases for human analysts. They spoke of vast subterranean operations under the noses of the American public, sometimes literally: one key pipeline involved the "tapping" of undersea data cables to harvest personal information on online activities. Over the next several months, journalists revealed that the National Security Agency (NSA) had spied on foreign diplomats and national leaders, that it had monitored players of online video games and even surveilled pornography consumption habits as blackmail fodder against "radicalizers." In a debate where critics of surveillance had been perennially marginalized as paranoid rabble-rousers, the files were credited as being the "first concrete piece of evidence exposing dragnet domestic surveillance."[8] William Binney, a former NSA employee who had told the public much the same things Snowden did with much less impact, thought that the material documents made the difference: he regretted that he did not take any himself, the "hard evidence [that] would have been invaluable."[9]

Yet the stream of revelations also provoked a great mystery: Just how many documents were there, how many secrets to be told? The exact size, scope, and location of the Snowden files captivated the American news media—as if getting it right would provide some handle on the knowledge on offer.[10] Not that anyone could figure out just how many documents even existed. Snowden himself never deigned to supply a number. In an interview with the German public broadcaster ARD, Glenn Greenwald claimed that he possessed a "full set" of nine to ten thousand top-secret documents;[11] a year later, he would appear on New Zealand television and speak of "hundreds of thousands" of documents.[12] Meanwhile, the US government also tossed numbers into the air. A Defense Intelligence Agency report to Congress claimed that Snowden took nine hundred thousand files from the Department of Defense alone, distinct from his haul from the NSA.[13] One of the most widely cited estimates claimed that Snowden "touched" 1.7 million files while contracted for NSA work in Hawaii[14]—a figure often misconstrued as documents *taken*.[15] The wider public, without any means to check for themselves, could only watch.

This seemingly trivial mystery around the numbers danced around a more crucial question: How can the files speak the truth about data-driven surveillance? How can the public know such complex, secret, vast technological systems? The files are a clandestine archive of documents, offered as a map of another secret archive of surveillance data. It is data about data, information about information, and, like Borges's infamous map of the empire, made to be as large as the physical empire itself, the files replicate the problems of scale and comprehension surrounding state surveillance systems. The rapid expansion of electronic surveillance systems after September 11 required a massive boost in the NSA's funding, and a corresponding boom in internal hires, new infrastructure, and outsourcing contracts to the private military-industrial arm of the surveillance apparatus.[16] By 2010, the government itself lacked comprehensive and precise metrics for mapping its own surveillance apparatuses or estimating the overall costs of antiterrorism.[17] As if a parody of corporations "too big to fail," the landscape had become littered with big data too big to account for.

The Snowden files, then, were not self-evident forms of proof but a collective mobilization of belief in knowability relying on the *appearance* of numbers. Very much in Wittgenstein's tradition, Steven Jay Gould writes that "numbers suggest, constrain, and refute; they do not, by themselves, specify the content of scientific theories."[18] Distinct from the mathematical order that generates these numbers, their public appearance often produces an impression of calculability, a groundless ground we conventionally agree not to doubt. Quantification has long been a social technology.[19] Each presentation of numbers translates credibility across people and things, and more generally contributes to the evidentiary reputation of numbers as something to look for and seek assurance from. And once this trust is (slowly) won, the faith in quantification—that is, statistics and probability as a way of seeing the objective facts underlying every kind of situation—injects a mythological strand into what is advertised to be the triumph of cold, impersonal reason.[20]

This is not to fall back on a false consciousness argument, where modern subjects are tricked into believing in a sham objectivity. The seductiveness of numbers is an essential aspect of the public's ability to trust in numbers, and numbers' ability to stabilize social norms of factmaking. Popular "scientism"—the overblown faith that science alone produces absolutely certain truth about the world—has become a radicalization of the kind of trust that normal science asks of the lay public. In the same way, numbers and statistics often become ciphers for objective knowledge production presumed to be occurring backstage. Sheila Jasanoff retells the views of an American lawyer, who argued that the deluge of charts, tables, and figures in court cases risked becoming a strategy of *painting by numbers*: as judge and jury stare blankly into yet another mystifying graph, the totality of the numbers, their very inscrutability communicates a certain sense of objective authority.[21]

Evidence of a Secret

These affective and impressionistic uses of numbers do not merely stabilize dominant narratives. The vastness of the files, sketched with a numerical brush, also supports the flourishing of speculation: What is the information that we now "have" but still cannot access? What remains secret about that which is technically exposed, and what wider landscape of secrets does such exposure make visible? The files were so vast that even Snowden himself could not confirm if he had personally read all of the documents.[22] The gradual drip of new leaks (table 2.1) not only successfully kept the files in the news for months but also added up to a marathon of information ingestion that the public struggled to keep up with. Even in the first week of the leaks, a survey suggested that 50 percent of Americans followed the news on surveillance "not too closely" or "not at all closely."[23] Those who sought to read the files and know for themselves found a bewildering morass of information, often requiring a great deal of technical and institutional context to parse through terms, such as *selectors detasked*, or code names, such as Pinwale and Egotistical Giraffe.[24] These many mundane gaps between the promise of revelation and the messiness of information meant that the leaks served to generate speculation as much as it settled them.

Out in the public, the Snowden files had become an indefinite archive: credited as a source of transparency and public information but in practice as an amorphous stream of gradual revelations, whose elusiveness mirrored the secrecy of the very surveillance state it sought to expose. For Derrida, the archive is the desire for an origin, an origin-as-truth; its very form reflects the desire for an ultimately impossible dream of total containment and retrieval.[25] Evidence does not extinguish uncertainty but redirects it and refocuses it. It is only because the documents exist that the public can enter into speculation, indignation, skepticism—even if nobody can be quite sure of what is and is not in those documents: the halo of potential justifications and harms still to be uncovered, the bulk of the iceberg still submerged. In the world of supermassive databases and hyperobjective tech infrastructures, the archive fabricates a sense of knowability—not through acts of deliberate deception but by serving as a container of the desire for knowledge and control.[26] Whether the voluminous cache of the Snowden files or the enthusiastically embraced proliferation of "big" databases, these enormous archives become mobilized as a mystical embodiment of the truth out there—and of the hope that all these secrets, all these complexities, could be ordered, bounded, and accounted for.

TABLE 2.1. Cryptome's table of Snowden files leaked by *The Guardian* alone in the first few months of the affair

Number	Date	Title	Pages
	The Guardian		276
	27 February 2014	GCHQ Optic Nerve	3
21	16 January 2014	SMS Text Messages Exploit	8
20	9 December 2013	Spying on Games	2
18	18 November 2013	DSD-3G	6
19	1 November 2013	PRISM SSO SSO1 Slide SSO2 Slide	13
18	4 October 2013	Types of IAT Tor	9
17	4 October 2013	Egotistical Giraffe	20
16	4 October 2013	Tor Stinks	23
15	11 September 2013	NSA-Israel Spy	5
14	5 September 2013	BULLRUN	6
13	5 September 2013	SIGINT Enabling	3
12	5 September 2013	NSA classification guide	3
11	31 July 2013	Xkeyscore	32
10	27 June 2013	DoJ Memo on NSA	16
9	27 June 2013	Stellar Wind	51
8	21 June 2013	FISA Certification	25
7	20 June 2013	Minimization Exhibit A	9
6	20 June 2013	Minimization Exhibit B	9
5	16 June 2013	GCHQ G-20 Spying	4
4	8 June 2013	Boundless Informant FAQ	3
3	8 June 2013	Boundless Informant Slides	4
2	7 June 2013	PPD-20	18
1	5 June 2013	Verizon	4

Source: Re-created by the author from "42 Years for Snowden Docs Release, Free All Now," Cryptome, February 10, 2016, http://cryptome.org/2013/11/snowden-tally.htm.

These basic contours of the affair identify a paradox that I call *re-cessive*. On one hand, the Snowden files materialize the unknown. It promises direct contact with the depths of state secrecy and technological complexity. The language of exposure, leaks, and shedding light expresses the familiar trope of knowledge as illumination; the materiality of the files provides veridical guarantee that they bring undistorted fact and information into the sunlight—the "best of disinfectants," as Louis Brandeis said. On the other hand, this rare artifact from a secret place, brought to the public as a beacon of transparency, now compels citizens to journey into that still-strange world out there.[27] Like the hyperobjective images of climate change, these files let us glimpse at the tip of the iceberg and, in doing so, make the still invisible iceberg an unavoidable topic of discussion. If the public previously generated its decisions and opinions by tacitly accepting the unknowability of state surveillance (for instance, by having no particular opinion of it or by dismissing any criticism as conspiracy theory), then the Snowden files compel reasonable citizens to speculate and extrapolate—not just because the files present new information but precisely because the files tell us there is so much we do not know and that this unknown must now be a matter of concern.[28]

This performative, incomplete, speculative relationship between the Snowden files and state surveillance systems spell out the asymmetries of visibility and knowability that characterize systems of datafication as public matters of concern. As chapter 5 shows, the NSA protested that there were good reasons for its surveillance systems to be so secret and inscrutable; a popular counterargument against Snowden's leaks was that disclosing these technologies would allow terrorists to better evade them and, indeed, that Snowden's actions had put lives of agents at risk. (In Britain, a senior Home Office official asserted that the leaker had "blood on his hands"—even as Downing Street, on the same story, put it on record that there was no evidence the leaks had harmed anyone.) In effect, the public is asked to invest their rights and beliefs in a system of knowledge production that requires ordinary individuals to be maximally exposed and the system itself to be maximally concealed. Such a situation pressurizes the relationship between knowledge and uncertainty. The ideal of the informed public is confronted with both surveillance's inherent need for secrecy and what Bernard Harcourt has called "phenomenal opacity"[29] and Frank Pasquale, the "one-way mirror":[30] the ways in which big data technologies become resistant to everyday, experiential grasp. The growing ubiquity of data-driven decision-making across not just intelligence agencies but also local law enforcement, and their interoperability across private systems, such as CCTVs in stores or cameras installed in individual homes, exponentially increase the distance[31] between individuals and their data. In this context, the traditional reliance on the virtuous cycle of transparent information for an empowered public begins to lose their bearings.

Connecting the Dots

New tools have a way of breeding new abuses. Detailed logs of behaviours that I found tame—my Amazon purchases, my online comments . . . might someday be read in a hundred different ways by powers whose purposes I couldn't fathom now. They say you can quote the Bible to support almost any conceivable proposition, and I could only imagine the range of charges that selective looks at my data might render plausible.
—Walter Kirn, "If You're Not Paranoid, You're Crazy," *The Atlantic* (2015)

November, 2015. With the Snowden leaks still fresh on the mind, *The Atlantic* magazine advised that paranoia is the new normal.[32] As humans promiscuously supply all manner of personal data to electronic networks,[33] the machines, in turn, communicate and triangulate ceaselessly in a wireless hum. Social networks know you have been to Alcoholics Anonymous, Google and Facebook know you have been visiting porn websites,[34] and state surveillance systems suck in an unknown proportion of your emails, your Skype calls, and your internet banking records. *The Atlantic* piece concluded that paranoia was no longer a disorder but a "mode of cognition with an impressive track record of prescience." (Three years later, the public would be told that many smart devices *do* listen in on their users while dormant—and that in some cases, human analysts access those recordings for product improvement purposes.[35])

To try to know secret surveillance systems is to learn to perceive a certain cohesiveness, to rescue some sense of certainty out of the muck, to be able to connect the dots. The early twentieth-century psychiatrist Klaus Conrad called it *apophenia*: the tendency to identify meaningful patterns in random data. He pegged it to the acute stage of schizophrenia. Daniel Paul Schreber, the mythological "origin figure" of modern schizophrenia, was indeed greatly concerned with a complete and systematic order of meaningful truths, describing an *Aufschreibesystem*, an automated writing system that might perfectly represent his thoughts.[36] Conrad's neologism, assembled out of the Greek ἀπό (away, apart from) and φάνειν (to show, reveal), shares with the much older paranoia (παρά [besides] + νόος [mind]), a clear pathologization of this desire for order and meaning. What does it mean, then, to say that paranoia has become normal, a sensible and prudent response to the exigencies of the world around us?

I would like to pursue this charge of normalized paranoia not from a psychiatric or psychopathological viewpoint but an epistemological one. In effect, *The Atlantic*'s conclusion amounts to a recommendation that we fabricate more actively and aggressively than before—and that such a shift in the norms of factmaking is necessary to cope with a data-driven society. To be sure, suspicions about government surveillance, and, more generally, a state's tendency to abuse its powers, has long been a public secret: something that is generally known (or assumed) but rarely becomes officially articulated.[37] What objects such as the Snowden files do is bring those subterranean ways of seeing out into the open of public discourse. It is not that millions of individuals will specifically feel that the government is out to get them. The change occurs not at the layer of subjective experience but in the normative structure of epistemological expectations. The files' appearance as veridical objects provokes a renewed focus on surveillance's secrets; the public is presented with an urgent necessity for constructing meaning even—or especially—in the presence of unknowns.

The recessiveness of datafication thus encourages the "ruthlessly hermeneutic logic"[38] of a paranoid subject—the intensification of that search for a grid of intelligibility that, in varying degrees and shapes, is a feature of any regime of knowledge. There is an apocryphal story that some conspiracy theorists were rather put out when the Snowden leaks happened: now that their theories had been proved right, they would have to come up with some new ones! For a more concrete example, consider a post on Reddit's /r/conspiracy, a hangout for conspiracy peddlers (or, as the site itself puts it, "free thinkers"): "If it weren't for Edward Snowden conspiracy theories would still just be 'theories' . . . High five to the sane ones <3."[39] This slippage between conspiracy theories and "just theories" reflects the fragile social boundaries that demarcate what is and is not an acceptable way to fabricate explanations. To label undesirable, deviant, threatening modes of knowledge making "conspiratorial" is to engage in a "rhetoric of exclusion," where the very act of naming marks that discourse out as illegitimate.[40] One does not, after all, engage conspiracy theories seriously to refute their various claims but summarily dismisses them from being "possible candidates for truth."[41] "That's just crazy" is the mantra of foreclosure that refuses to enter into reasoned debate with the theory at hand. (The same way in which our parent had told the Wittgensteinian child, "Stop asking; just believe that this is a tree.") However, events such as Watergate or the Snowden affair push the pseudo-conspiratorial, semi-acknowledged truths about government surveillance into more respectable public discourse. Much maligned and yet widely circulated and entertained, conspiracy theories demonstrate the ways in which the candidacy to knowledge is strictly policed. At the same time, these disavowed rejects are constantly smuggled in to cope with looming uncertainties. Like paranoia as a structural, rather than a pathological, symptom, conspiracy theories reflect not an antimodern strain of irrationality in the system but a *useful* by-product of rational knowledge production.[42]

This shift in what sounds paranoid or appropriate is thus not restricted to card-carrying "free thinkers" but reprises what Richard Hofstadter called the paranoid style in American politics: a mainstream tradition of conspiratorial and indignant mode of expression that could be found in McCarthyist America of the 1950s or even the moral panic over the Illuminati in the late eighteenth century.[43] In particular, Hofstadter argues that the right-wing paranoia in his own time—the 1960s—is founded on a sense of presumptive dispossession: the idea that they have *already* lost the country to powerful and shadowy forces that control their every move. This postapocalyptic imagination provokes not only a militant reaction but a general sense of agency panic. Specific fears about communist plots or omniscient machines supply the broader sentiment that the liberal ideals of individual autonomy and freedom are under siege.[44] Mainstream news media coverage of the Snowden affair was awash with conspiratorial language, especially among those critical of the whistle-blower. Speculations that Snowden was a Russian or Chinese double agent, or at least their gullible puppet, were fuelled by Keith Alexander and other high-ranking NSA officials.[45] One *Washington Post* piece suggested that Glenn Greenwald, Julian Assange, and others had conned the gullible Snowden into risking his life for the former's ambitions—at least, before the paper had to issue a series of corrections to dial it down.[46]

The Snowden files became generative of new theories, new speculations, projecting ever larger shadows behind the actual facts it revealed. What matters is not just the information these documents provide but a variant of what Tor Nørretranders calls exformation: the bits of a message that are "explicitly and knowingly discarded,"[47] the bits that the available information leaves unsaid and unproved but that now gain a social presence in a provisional and anticipatory form. A paranoid epistemology is thus an apophenic one: the trouble is not that meaning is secret, hidden, or lost but that it is too much and everywhere.[48] Yet to label such strategies irrational would be to reproduce the ideal notion that information should lead us to proof and certainty. Instead, we might look to what Tobin Siebers called the "Cold War effect": a generalized epistemological climate where paranoia and suspicion were seen not as delusions or pathologies but as virtues, and to be paranoid was not to be ill but to be in tune with contemporary reality.[49] Indeed, Cold War rhetoric was frequently reprised in a concealed form in Snowden-era paranoia.[50]

Merleau-Ponty understood that the "mad" experience their own madness as no error or illusion but a naturalized and intuitive access to truth. A schizophrenic experiences voices not as hallucinations superimposed over reality but something as genuine as the ground beneath our feet. (Thus, Merleau-Ponty describes a schizophrenic woman who believes two individuals with similar-looking faces *must* know each other: a connection that "normal" humans would dismiss as apophenia gone haywire, but for the woman, this is simply common sense.[51]) The point is that any given system for rendering the world around us into intelligible pieces requires some reliance on presumptions about the unknown—a reliance that, to outsiders, appears arbitrary or nonsensical. It may be technically prudent to wait until all the facts are in hand, but in the case of a secretive surveillance program and the logic of preventive prediction, nobody will ever reach such a privileged position. The public, as much as politicians and counterterrorism officials, are increasingly asked to judge and act well in advance.

There were cautionary voices, encouraging the public to return to a more conservative range for crafting explanations out of available data. Some pointed out that the risk posed by terrorist attacks remained rather small compared to, say, gun shootings.[52] Others simply insisted that criticizing surveillance programs would require presuming too much corruption and impropriety on the NSA's part for it to be realistic: "fearing the NSA . . . requires you to believe that hundreds, if not thousands, of American employees in the organisation are in on a conspiracy."[53] The only reasonable solution would be to trust in the NSA because not trusting would require us to be, well, paranoid. These disputes reflect the contested recalibration of what counts as *reasonable*, of what might count as a conventionally acceptable performance of reason between paranoia and naivety. Here we are reminded of a basic lesson in machine learning around overfitting and underfitting. Simply put, analysts are instructed to avoid following the data too closely, resulting in a model that reflects the vagaries of the available data rather than the underlying phenomenon, or not closely enough, in which case the result fails to properly model the trends in the data. Whether a model is appropriately fit thus is a question of human judgment, a convention guided by circumstance as well as mathematics. Even as these technical practices were being challenged as full of error, uncertainty, and arbitrary judgment, the human debate around these technologies was facing a similar dilemma: What counts as a "reasonable" response to the asymmetric information environment of the Snowden affair?

It was a question with direct relevance to not only the public deliberation but also in institutionalized decisions around known and unknown—such as the courts. Snowden's first leaks in 2013, and the preceding leaks by *The New York Times* and *USA Today* in 2005–2006, precipitated a series of legal cases against government surveillance. In each of these, the most important issue turned out to be a basic question of available facts: What kind of harm is *known* to be caused by surveillance? Despite the new availability of Snowden's files, efforts to contest NSA surveillance at the judicial level struggled to gain standing due to the difficulty in constructing a definition of surveillance harm that is compatible with the existing legal conceptualization (in the United States) of harm as "concrete, particularised and actual."[54] In *ACLU v. NSA* (2007), the district court concurred that phone/internet data collection is both unconstitutional *and* counts as concrete, particularized, and factual harm; however, the Sixth Circuit Court of Appeals ruled that the injury claimed is "mere belief" of intercepted communications, and the lack of any "personal" harm, only a "possibility," denied them standing.[55] A similar reliance on a narrow definition of harm has also dogged efforts to sue technology companies for breaches of data privacy.[56] Such debates reflect a fundamental problem with public secrets: What must one "know" to bring the unknown to trial? What should and should not count as "known" in the face of such relentless uncertainty? The changing standards of reasonable extrapolation thus correspond directly to the legal and institutional scope for recognizing and addressing datafication's consequences—a problem we shall return to in chapter 5.

The Transparency Illusion

This entanglement of knowledge and uncertainty makes a parody of the contemporary enthusiasm for transparency. Transparency is axiomatic for whistle-blowers, and Snowden, too, framed his actions in this light.[57] More generally, the concept had grown in prominence over the preceding decades, empowered and idealized as a universal tonic for liberal democracy and the Enlightenment.[58] Since the 1990s, buzzwords bloomed by the dozen in the wake of enthusiasm about the transformative powers of internet communication technologies: e-government, e-transparency, e-democracy . . . as if digital technologies would finally eradicate ignorance and misinformation and furnish the optimal basis for the public's rational judgment.

Such mythologization of transparency as an unalloyed good and universal solution reflects two kinds of conflations about how knowledge works in the data-driven society. First, this idealized belief in transparency involves a "virtuous chain": the public is injected with information, which is linearly correlated with more rational deliberation, and, in turn, the arrival at an "optimal" decision.[59] Like the aforementioned pyramid from data to wisdom, this consolidated, linear model equates transparency with a global good, sweeping away the long essential role that secrecy and opacity had played in Western statecraft.[60] We find here another instance of the fantasy of epistemic purity, one that stands blissfully ignorant of what politics *is*. As Latour quipped, asking politics to tell unvarnished facts without rhetorical trickery is like asking science to tell truth without peer review, without experiments—and, yes, without any mediation of its own![61] Second, and related, is the belief in transparency as an indispensable cog in the apparatus of liberal democracy. Kant's *Sapere aude!* here becomes a directive for stuffing each and every citizen with maximum information about issues of public import. Yet, as we have seen, there is no easy connection between the theoretical availability of information and its uptake as knowledge.[62] As with the Snowden files, the presentation of solid, reliable information can *increase* the public labor of speculation and inquiry until citizens simply cannot keep up.

What becomes clear is that transparency is not a binary opposite to secrecy, the purifying sunlight idealized by Louis Brandeis. It is instead part of a wider ecosystem of knowledge that allows the circulation of ideas and impressions across different types of truth—types that exhibit different gradations of openness and publicity. This system might involve formal and institutional moves, such as declassification of formerly secret documents. It also includes perceptual and social shifts in which a public secret becomes a matter of concern or a percolating suspicion becomes legitimized into a belief that citizens feel they may wear on their sleeves. Importantly, these practices are not arrayed in a linear scale of progressive visibility or informed public deliberation. Consider electronic state surveillance's pre-Snowden status as an open secret, in which the public suspects and even assumes it is happening, but an official game of denial just about maintains the technical status of secrecy. As one reading of Kant's secrecy suggests, "the veil always also unveils, or promises an unveiling, but that promise, and the prospect of finally seeing what is behind it, are also part of the veiling."[63] Although transparency presents itself as a necessary harbinger of truth, it does so precisely by idealizing a specific conflation of publicity, honesty, and innocence—and forgetting the myriad other ways in which claims to knowledge may be paired up with speculation, interpretation, and judgment.

The genre of media *exposure*—which also became the basic format for reportage in the Snowden affair—also adds its own patterns to the unveiling. It is not coincidental that, at least in the American case, transparency emerged as a universal virtue in tandem with the rising centrality of exposure in journalism.[64] Both the sensationalist tabloid exposé and the somber investigative report share this foundational assumption that there always remain more secrets to be uncovered, that each story, each leak, gives an approximation of the rest of the iceberg submerged beneath the visible. Likewise, the frenetic pattern of constant updates in new media platforms[65] cultivates the "public's persistent feeling that 'there is always something'" more behind the scenes.[66] After all, no exposure can ever claim to reveal the whole truth, nor can it guard that truth from the swarming multiplicity of interpretation. As such, this normalized expectation of exposure is met by a prevalence of cynicism (in Sloterdijk's sense[67]): the revelation invites not acceptance but further interrogation of the leaker and the leaked, generating an economy of speculation that feeds on each effort at transparency (or, for that matter, secrecy). Transparency's practical function, then, is a clearinghouse, a switchboard: a technique that redraws the local boundaries of what counts as speculation, what counts as "on the ground" facts, what may pass as consensually assumed truths. Brandeis's sunlight receives a McLuhanian correction: illumination is neither natural nor neutral but a technological medium.[68]

The relation between transparency's idyllic promise and its multifaceted practical function can be better understood when we remember the highly contingent—and recent—history of its emergence. As Michael Schudson has shown, today's ubiquitous celebration of transparency only took off in the United States during the mid-twentieth century. It did so not through a broad public demand to "know for itself" but through political shifts in relations of trust and communication across the branches of government and media industries, such as a more adversarial model of journalism and the rise of public advocacy groups.[69] Inaccurate accreditations—such as the belief that Thomas Jefferson called information the "currency of democracy" (it was, in fact, Ralph Nader)—bestow mythical origin stories to what is in reality a more profane and youthful idea. This recognition of transparency's historicity forces a new perception of its present form: not as a fundamental ideal for the fulfillment of deliberative democracy but as part of a specific generation of (imperfect) machinery for that deliberation. Fast-forwarded to the times of vast electronic surveillance systems and their subjection to digitally proliferating leakage, transparency constitutes not an external panacea to these problems but their companion in mediating what kind of veridical force is given unto that which we think we know.

The Burden of Knowing

If transparency is a switchboard for different ways of knowing, each marked by the kinds of decisions and interpretations they authorize, then we must inquire into the practical consequences of fetishizing transparency. What kinds of powers and responsibilities are given over to the public in an act of transparency? The Snowden affair is one example in the wider story where the Enlightenment injunction to "know for oneself" thrusts an impossible labor onto the internet-age citizen. In the context of liberal, representative democratic societies, transparency mobilizes the citizen anew with an old responsibility: not just to participate in politics in prescribed moments and ways (e.g., voting every four years) but also to become an unblinking eye poring over every aspect of government. The citizen has been recruited as a free auditor for the state. This is to be distinguished from earlier forms of citizen redress, such as petitions of grievances and injustices. The long Western history of petitions, from written pleas to the Roman emperor to the *cahiers de doléances* in eighteenth-century France, was not the normal duty of subjects but extraordinary actions—and the work of assessment and redress remained the task of the governing prince.[70] This case was also for the literary trope of the king who speaks with his subjects in disguise to hear their grievances, most famously Shakespeare's Henry V and James V of Scotland's legend as "King of the Commons."[71] Again, it remained the king who must listen, gather data, make his population legible, and reconfigure his apparatuses of government according to that knowledge. In the e-transparency paradigm, however, the government (or the whistle-blower) merely uploads, makes "available"—a passive position, after which it is the public's responsibility to request, read, cross-reference, judge, and prosecute. The proof may be *in* the Snowden files, but the burden of proof is *on* the subject.

The problem is that much of the time, it is a burden that the members of the public cannot afford to or are reluctant to bear.[72] When another realm of online surveillance—corporate data mining—became scrutinized for invasions of privacy, one popular solution was to push for greater transparency on the part of online platforms. Predictably, the result was an even greater onslaught of privacy policies that many people do not want to read, do not have the time to read, and do not have the background knowledge to fully understand. As one study showed, it would cost 781 billion USD per annum in salary if Americans used their working hours to read the privacy policy of every website they visited.[73] Well intentioned as they may be, such measures risk drowning the citizen in pointless information. And so, the impossibility of fully taking up, or "owning," the burden of transparency produces a new chain of deferrals and delegations. Set against systems for the production, circulation, and resale of information that are too distributed, complex, and technologically backgrounded for human upkeep, the tacit ideal of the maximally informed subject summons an overbearing specter of guilt. Although maintaining a skeuomorphic appearance of a liberal public sphere, digital transparency becomes an extension of the entrepreneurial, individualized responsibility that we have sloganized as "neoliberal." Even as technology promises that information shall be free, citizens are asked to work for free to support these growing mechanisms of truth production. Here, transparency functions as a false dawn, or even a barrier, to becoming political.

What if we thought of the work of politics, the work of being informed, as a form of labor? In economic terms, transparency appears as a practice of outsourcing, of creating externalities: costs that are not counted by the producers directly but are passed onto the rest of society.[74] The fantasy of e-government relies on this standing reserve of public engagement that transparency shall mobilize for free. Indeed, such mobilization already occurs in the American tradition of citizen surveillance: from vigilante neighborhood watches to the use of social media by police to receive tip-offs, the state has long relied on ordinary subjects' sense of autonomy and agency to supplement the work of government.[75] Closer to home, the subject of the data-driven society is already well trained in another kind of free labor—the work of staying connected to keep uploading photographs, to keep participating—that generates the economic surplus of platform capitalism.[76] In their capacity as citizens, those same subjects are enjoined to stay more informed about more things than ever—as a way not simply to empower the good liberal subject for the demands of a complex information society but also to defray its costs.

The moralization of transparency has pernicious effects on the ideal of the public that "knows for itself"—effects that recall the earlier warnings from writers such as Walter Lippmann. In a world where information encourages speculation as much as consensus, transparency is too often a Trojan horse, not a panacea. Again, there are uncanny parallels with what we have said of conspiracy theory. If the concept of conspiracy taints the information thus labeled and expels it from the normative realms of deliberation (even as it continues to circulate and communicate), the name "transparency" invokes the presumption that a full and equal distribution of information is possible and desirable. If the shining light of novelty blinded early internet-age optimists into believing that everyone really could become the public that knows and decides for themselves, then we are still struggling to clear the afterglow from our eyes. To know through deferred and simulated means, to agree tacitly to exclude certain doubts or uncertainties from debate, and even to operate within restricted information flows is to protect the possibility of consensus and shared grounding in a democratic society. The untrammeled pursuit of transparency opens each time a hermeneutical Pandora's box, even as it promises to illuminate and disinfect the black box of datafication.

Mary Douglas once suggested that "certainty is not a mood, or a feeling, it is an institution"; that is, "certainty is only possible because doubt is blocked institutionally."[77] In other words, it is the product of conventional norms that we learn to avoid the stigma of conspiracy, the abyss of paranoia, and exercise our public judgment on the basis of what may be officially admitted (and what is unofficially and tacitly understood). We learn not to question Wittgenstein's subject and to operate on the shared basis that what I say I know to be a tree is indeed a tree. The corollary is that uncertainty is not kept at bay by the sheer strength of our knowledge and reason but by the decisions we make on what to believe and how to believe. And when those decisions become challenged by changing conditions, such as the vast and backgrounded complexity of new technological systems, the rules governing those boundaries begin to shift. The moral question, then, is clear: What kinds of boundaries and tacit norms should we adopt in an age of excessive information, of ever more ubiquitous yet concealed technological systems, and of unparalleled speculation in the public domain?

* * *

In the winter of 1998, the philosopher Thomas Nagel published "Concealment and Exposure."[78] It asks: Should information always be analyzed, disseminated, acted on? In what cases can new information be distracting or inappropriate to the judgment at hand? Written amid the scandal over Bill Clinton's extramarital affairs and his attempted impeachment, the question applies far more broadly to the benefits and limits of transparency. Nagel understood that information is not always beneficial in the same way and that it can infect public discourse with a cacophony of the trivial, the irrelevant, and the half-true. He argued that the increasing pressures for transparency need to be balanced by a corresponding provision of tolerance and *nonacknowledgment*: to know something and to not speak of it, to not bring it into one's decision-making. Since Nagel wrote his piece, such balance has only broken down further. Whereas Clinton was almost removed from office over his adultery, Barack Obama, the next Democratic president, was subject to incessant accusations about his religious allegiances and even his birth certificate. The question of what *should* be relevant to a given judgment was overwhelmed by transparency's slogan that everything that can be scrutinized should be. The argument for nonacknowledgment exposes the unbalanced nature of transparency as a style of fabrication and its dangerous proximity to political cynicism.

Perhaps the most counterintuitive aspect of Nagel's argument is that we should use nonacknowledgment to exclude the kinds of information about which *we know the public cannot come into agreement.* "Leave people to their mutual incomprehension," Nagel advises: pick your battles or risk devolvement into interminable squabbles over each citizen's allegiance on every kind of issue. His chosen example is the generic demand that citizens "stand up and be counted"—perhaps by reciting a patriotic slogan or by professing their support for multiculturalism. What was meant to be an ethical and reflexive move of disclaiming one's bias comes to support an indiscriminate demand for transparency. In the Snowden affair, this exhibitionist tendency demands that every actor plot themselves on a binary grid: hang on, before you say anything— which side are you on? Do you believe Edward Snowden a hero or a traitor? Which side are you on in this war between regimes of truth? He is a hero, said John Cassidy of *The New Yorker*, Shami Chakrabarti of *The Guardian*, and civil rights groups such as Amnesty International;[79] a traitor, argued Fred Fleitz at the conservative-leaning *National Review* and politicians such as then former vice president Dick Cheney.[80] Some, like Nate Fick writing for *The Washington Post*,[81] decided to sit on the fence and say a "little bit of both." Yet such insistence on disclosure lends itself to prejudiced readings of those actors' discourse. It reflects not the opposite of having good faith in other members of the public but the very *lack* of good faith. Ironically, this exhibitionism erodes a useful fiction central to the "virtuous chain" of transparency: the idea that the public will judge each argument in a fair and reasonable way, making proper use of available information to reach the optimal decision.

However, Nagel's analysis is blind in one important respect: it presumes that consensus is possible as long as codes of civility and nonacknowledgment invisibilize intractable differences. This blind spot is all too similar to the way the early Habermasian public sphere was often idealized as an open space for rational deliberation. Scholars such as Nancy Fraser have shown at length how such inclusivity and equality were often restricted to a small group of citizens—often white male bourgeoisie who read and wrote for each other. In this sense, nonacknowledgment risks reproducing the boundary policing work we have seen in the definition of conspiracy theories. Especially telling in this regard is Nagel's example of sexual thoughts. Woman D applies for an academic job in C's department, who is "transfixed by D's beautiful breasts." Yet C refrains the best he can from expressing his "admiration," and D accordingly refrains from voicing her disgust.[82] Here, nonacknowledgment hardly solves the problem. Even if we very generously interpret C's behavior as that of a polite fellow who does his best not to objectify D, the result is neither equitable nor desirable. Conventions such as civility and nonacknowledgment secure knowability precisely by sacrificing whatever does not fit. D's ability to contest her objectification is curtailed by norms of nonacknowledgment—especially because C, in his position of power, is likely to exert greater influence over what is or is not suitable for disagreement. Across counterterrorism operations and quantified analyses of individual bodies, we will continue to find these political and ethical implications of the shifting boundary between known and unknown, the sayable and the inadmissible.

In short, the idealization of transparency risks conflating exposure with truth and expression with honesty. In doing so, it encourages speculation of a promiscuous kind—one that erodes and overwrites existing norms for the boundaries of relevance and credibility. There is a telling parallel here between the exigencies of big data technologies and the challenges facing the public in a data-driven society. If the former involves enormous quantities of data processed by automated machines, leaving users struggling to figure out how to make sense of it all, the latter asks the public to "know for itself" despite being ill equipped to consume this information responsibly and effectively. The relation between the injunction to know, excessive information, and speculative uncertainty occurs not only in the public's effort to know *about* state surveillance systems but also in the state's efforts to know *through* those systems also. The next chapter turns to this latter side of the problem, understood through another kind of fabricated object: the figure of the "lone wolf" terrorist.

The Gap

In Agatha Christie's novels, we find a trope of revelation: when enough "secrets" (i.e., objective facts) have been accumulated, the illusions topple all at once to reveal a perfect picture of the crime. The pleasure of this revelation is itself an expression of our shared intuition that, back in real life, things rarely seem to work out so neatly. Sherlock Holmes, too, insisted on a progressive and ultimately conclusive process: "when you have eliminated the impossible whatever remains, *however improbable*, must be the truth."[83] Holmes's world, of course, is a conveniently finite and localized one. It is rare that the suspects do not wear every relevant aspect of their psychology and history on their person for the discerning eye of the detective. But what happens when tens of thousands of government-employed analysts roam the four corners of the internet, from massive headquarters the size of a small city? (The NSA's Fort Meade is larger than Cambridge, Massachusetts, in land area.) What happens when the nature of data collection mechanisms is such that nobody, not even the collectors, knows whether your data will ever be seen by a human? The linear eradication of the secret is replaced by an open struggle of speculative hypotheses that must all admit their partiality and uncertainty, even as they bid publicly for our belief.

This entanglement of knowledge and uncertainty comes down to a *gap* between the document as evidentiary object and the "knowing" it is meant to produce. It defies the transmissional imagination that proving, verifying, and informing humans can work like a digital file transfer. This gap is at the level of neither metaphysics nor the content of individual experience but the embodied and social structures that any regime of knowledge depends on. Known and unknown, transparency and secrecy, turn out very rarely to manifest in such pure forms. The Snowden files, celebrated and feared in equal measure, were supposed to provide truly solid, material grounding, as solid as it gets short of catching an NSA agent nibbling at your Ethernet cable. But the documents end up bringing in the distant and black-boxed "out there" into public concern. What does it mean for an object to acquire the status of proof? What other proof must exist for this object to tell its truth, and what are the subterranean beliefs, objects, conventions, and rhetoric that prop up its veridical authority? The recessivity of data and technology, so fundamental to surveillance's project of knowing, undergirds these phenomena.

3

Recessive Objects

In *The Watchers,* the journalist Shane Harris tells the story of the "BAG":[1]

> [It] stood for something unexpected: Big Ass Graph. In the late 1990's the engineers and systems gurus at the NSA became enamoured of computerised graphs to display huge sets of information . . . The graph builders of the NSA wanted to turn raw data into visual knowledge.
>
> But if the BAG was a useful tool, it was also a demanding one. For the BAG to tell them things, the [terrorist] hunters had to fill it . . . the resulting analysis overwhelmed them. The BAG's very design, the way it compressed information into more manageable forms, actually diluted nuance . . . For [the BAG] to tell them things, they had to feed it. But the more they fed it, the less it actually told them.

The big-ass graph materializes the gap between the human subject and the world out there, parallel to the problem of public knowledge in the Snowden affair. Deep within the hyperobject that is the surveillance apparatus, its human agents struggle to come to grips with a hyperobject of their own: an increasingly unpredictable and distributed terrorist threat. Between at least 2006 and 2013, the agency's internal mail service distributed weekly columns from the "SIGINT [Signals Intelligence] Philosopher." They contained brief musings that suggested "data is not intelligence" or that analysts increasingly face "analysis paralysis" and "cognitive overflow."[2] Yet the massive expansion of data collection and storage continued, under the idea that if everything could be tracked about everybody, the hidden correlations to the most unpredictable threats could be disclosed. The database as archive thus reprises Borges's famous story of the Library of Babel: a place containing every book ever written, every book that it is *possible* to ever write.[3] Initially celebrated as a holy grail of knowledge, its denizens quickly find that they are stuck in a constant state of limbo, where the answer is always theoretically available—if one were to just find the right book.

Chapter 2 described the Snowden files as a recessive object. The files promise to extend our knowability but, in doing so, bring into affective and discursive presence the uncertainty that they can never quite eradicate. They serve as reliable forms of proof and evidence but, by the same token, serve as catalysts for speculation and doubt. This chapter argues that recessive objects emerge from the contradiction between the groundless ground—the ultimately unprovable and conventional nature of our claims to knowledge—and datafication's promise of objective knowledge. In the process, the human subject is caught in the gap between the expectation to know for itself and the asymmetric and opaque forms of technological factmaking.

Recessive relations are, in a basic form, inherent in any knowledge claim: when Wittgenstein's subjects say, "I know this is a tree," they are setting the limits of our horizon as well as illuminating our path. But specific configurations of recessivity have distinctly political and ethical implications. Which forms of media, invested with what kinds of commercial and political interests, are elevated to the role of translating between knowledge and uncertainty—resulting in what new relations of dependency? Just as the printing press catalyzed a large and distributed reading public and radio and mass media provoked warnings of political paralysis and "narcotizing dysfunction,"[4] big data and smart machines introduce new formulas of asymmetric visibility: since we cannot handle the voluminous data processed by the machine, it is impossible for us to fully justify its decisions to you. Recessivity must be understood as a process of *selection*: What becomes classified as possible, available, expressible, and debatable while sending others to the groundless ground or beyond the horizon of the knowable?

These ambiguities become particularly pronounced through the data-driven society's constant demand that we know, judge, act on things out there beyond the horizon of our experience and sensibility. Analogous to the myth of transparency as sunlight, the data hunger of the big-ass graph expresses a desire for saturating the horizon, for complete capture of the world of possibilities through technology. This is no simple contrast between an ignorant public and the informed experts. Police personnel, for instance, might produce and process data themselves but only through third-party software supplied by private contractors wielding analytical mechanisms that they cannot fully understand.[5] Accordingly, this chapter traces the life of another recessive object, this time central to the state's own effort to know the world through data: the figure of the lone wolf terrorist, set against the narrative of terrorism's growing unpredictability.

Receding Horizons

To say something "recedes" typically implies an embodied experience of perception: when something recedes beyond our horizon, it diminishes, becomes distant, goes beyond our grasp. It recalls the most basic question of phenomenology: How do things in the world appear to our senses, and disclose their meaning? In Husserl, this fundamental relation becomes expressed as *presence*: the basic quality necessary for any thing to be sensible to the human subject. Here, becoming sensible involves not simply the mechanistic detection of stimuli but its recognition as something comprehensible—a process that is not reducible to language, symbols, or higher cognition. Presence is thus described as "what meaning cannot convey": a truth that happens (*ein Geschehen*), not information to be input and processed.[6] Typically, such presence is submerged beneath conscious reflection.[7] Data-driven knowledge relies on excavating and rewiring this architecture of the sensible insofar as it seeks to extend the reach of the human sensorium and deliver the kind of data that it could not have otherwise collected. The technology encourages a distributed kind of phenomenology, where we actively feel that what we know, what we see, and what we feel is not quite "our own."[8] Recessive objects such as the Snowden files pull human subjects out of sync with their own phenomenological horizon, compelling us to contend with knowledge, opinions, and feelings that require leaping across the gap—whether through the myth of transparency as disinfectant or through the normalization of paranoia to connect the dots.

Technology, of course, has a long tradition of aspiring to an extension of the human—an extension which is too often fantasized as seamlessly empowering. But the ambiguity of recessivity results in not a simple extension of the knowable but a proliferation of absent presence. To say absent is not to imply a simple inverse or lack of presence. Absence itself can be noticed, perceived, and made into a collective object of concern.[9] When the Snowden files point toward the deep secrecy of the state surveillance apparatus, the latter becomes caught in a position that is neither simple presence nor absence. On one hand, the files become stand-ins for the absent truth, bringing it out of the forgotten category of "unknown unknowns"; on the other hand, they cannot bring the secrets entirely to light and serve to constantly remind us of that absence. Hence, the files work to encourage, rather than extinguish, conspiratorial and paranoid forms of speculation. In her reading of Heidegger, Sianne Ngai points out that moods (*Stimmung*)—and especially the mood of anxiety—provokes a sensitivity for the "existential structure" of things.[10] The collective anxiety surrounding surveillance and terrorism has us constantly brush up against this amorphous boundary between the known and the unknown, the leak and the secret, the visible and the invisible.

Although recessive objects may not quite deliver on certainty, they do possess clear practical utility. They serve as our equipment[11] for extending and redrawing the felt boundaries of what can be known—that is, what we can reach out and feel able to connect with, control, influence, and understand. This sense of contact is defined not by the brain's accumulation of information in a machinic metaphor but a sense of connectivity that is affirmed through the rhythm of bodily experience: the indefinite unrolling of the Twitter feed, the regular ping of exercise metrics from the wearable wristband. In this context, the subject need not believe in data's promise of better knowledge literally and entirely; they need only participate and enjoy the immediate benefits. The more things, discourses, heuristics, and institutions we wrap ourselves around with, the more we can cheat the limits of our horizons and achieve a more expansive, if distributed, kind of agency.

This recessive bargain requires, however, a certain reconfiguration of the good liberal subject. In the face of so many uncertainties and techniques of proof that seem to build on rather than extinguish them, one can hardly insist that the individual subject "knows" all the facts and exercise their reason. What matters is the constellation of people, things, and processes that "know" on my behalf and what kinds of responsibilities, decisions, and rights that effectively leaves for me. (This dilemma can be found even in Kant's early formulation of the Enlightenment, where the tension between the overcoming of authority and the necessity of expertise is never entirely resolved.[12]) In this context, the proliferation of sensors, measurements, and data has given rise to a range of posthumanist ideas about how human experience is "engineered": not merely augmented or supported but also shaped and generated prior to human reflection or action. Mark Hansen argues that "twenty-first-century media" are distinct because their engineering of human experience entirely bypasses, occurs prior to, and in sensory regions inaccessible by the human subject.[13] Here, media technologies are not merely tools enacting the intentions of present humans; neither are they limited to augmenting human action and the senses. These technical objects observe, collect data on, and indeed "sense" the world at a level that human subjects have no access to. As chapter 4 analyzes in detail, the special epistemic benefits of self-surveillance technologies are founded on this nonhuman difference, where the data is valuable precisely because it lies beyond the human senses and its insights are accurate precisely because they bypass human cognition. Such machinic sensibility leads to new mediative structures for managing this informational surplus.

This situation results in, Hansen argues, a "dispersed, environmental, non-subject-centered subjectivity,"[14] where conscious experience is always already the product of infiltration by technical engineering. Variations on this theme have been floated by other scholars, often in terms of technical objects' capacity for qualified forms of agency and cognition.[15] The takeaway is that technology is not merely an extension of man but also produces and operates *inhuman* processes.[16] Recessivity is the bargain of knowing but not knowing for myself, sensing but not sensing for myself. What begins as a phenomenological question—how we are "extended" beyond our horizons to know the world beyond—leads to a suite of political and moral problems over who (or what) knows on my behalf and how much say I have over that process.

Data Hunger

As we have seen in chapter 2, the Snowden files combined an excessive knowability (and the moral injunction to engage with such expansive knowledge) with the withdrawal of knowing processes into hyperobjective systems. Crucially, this process was not limited to the public-facing side of the affair. If it were, we might construe the problem as one of educating the public and the mass dissemination of messages—founded in the belief that expert insiders do really know "for themselves." Instead, the recessive problem in the public's knowing about technology and state secrets finds a parallel in the state's knowing *through* technology the dangers of global terrorism. The rapid expansion of data-driven surveillance, catalyzed by the September 11 attacks, involved the legitimation of a wide-ranging data hunger that posited the indifferent and recombinant logic of big data analytics as the answer to the uncertainties of terrorism. Parallel to the recessivity of the Snowden files, it is exactly surveillance's promise to codify and calculate danger through the enormous overproduction of data that becomes a breeding ground for new techniques of speculation.

During the 1990s, the NSA was making its case to rest of the government that terrorism had become radically unpredictable and diffuse and that this qualitative transformation of the danger faced by America required a new paradigm in surveillance as well. The September 11 attacks would finally provide this narrative with extraordinary political momentum. A 2002 Joint Inquiry Staff statement, having analyzed counterterrorism efforts in the years leading up to 9/11, argued that a "new breed of terrorists practicing a new form of terrorism" was afoot, evolving from state-supported, limited (targeted) casualty attacks to stateless, flexible, more secret, more meticulously planned, indiscriminately high-casualty attacks.[17] The statement was endorsing line of analysis already found in earlier reports such as the US State Department's Country Reports on Terrorism.[18] In the public domain, the 9/11 Commission's report helped popularise the narrative that the nation suffered this traumatic attack because it could not "connect the dots" on the information it already possessed.[19] Such debriefings helped present a sense of necessity for surveillance that was more comprehensive, more powerful, more, more, more. The rise of twenty-first-century dragnet surveillance was thus predicated on the identification of a new breed of terrorism—even as this narrative reprised earlier fears about the infiltration of communist threats in the mid-twentieth century.[20]

The results of this new narrative were material and tangible. First came the ill-fated "Total Information Awareness" [TIA] program, led by John Poindexter—infamous for his role in the Iran–Contra affair, rehabilitated to lead the newly established Information Awareness Office. Contrary to the later secrecy of the NSA, Poindexter held a public briefing to explain that although "there will always be uncertainty and ambiguity in trying to understand what is being planned,"[21] the terrorist's "unique transaction signature" could be extracted through techniques such as biometrics analytics (HID), semantic analysis of text (TIDES), and speech-to-text conversion for collecting and analyzing auditory communications (EARS). Metadata surveillance was not a luxury; it was just what you had to do to keep up with the reality out there.[22] Ironically, Poindexter's attempt at relative transparency backfired, not the least because of his notoriety. Following public outcry, Congress withdrew funding for the new office in 2003, and Poindexter would once again retire from public service. However, many of its programs would later be discreetly re-created and extended at the NSA, which would, over time, become the primary agency for mass electronic surveillance programs. The NSA's director during this time, Keith Alexander (2005–2014), argued that the agency's job is not just to look for the needle in the haystack but to "collect the whole haystack"[23]—a reasoning replicated in internal NSA communications. Poindexter's vision of having "all the dots" would endure as a guiding vision for state surveillance operations to come.

The "whole haystack" approach also replicated the patterns of indifference and recombination in big data analytics. First, this data hunger was more indifferent, and thus more inclusive, about the data it collects in comparison to traditional methods. Because you do not know what you are looking for until you already have it, you must always have, standing in reserve, everything you can afford to collect.[24] Second, it blurs the boundaries between civil and military, domestic and foreign, innocent and guilty. One of the most controversial aspects of the Snowden leaks was the risk of "incidental" collection: that in seeking to target a foreign individual suspected of terrorism, the NSA might also collect information about Americans (e.g., because they are copied in an email to the target). For instance, the NSA's Section 702–authorized collection of email communications is strictly limited to "domestic" collection. However, its techniques target not human individuals but electronic "selectors," such as IP addresses or email addresses, that will often involve transnational data flows and multiple human participants. The data must often first be collected indifferently and then "minimized" through cleaning and redaction on the basis of what is legally permitted and analytically useful. Here, the collection of the whole haystack reflects not only a technical problem but an ontological attitude as well. Given the post-9/11 presumption that anybody could be a potential terrorist, there can be no hard boundary between innocent and guilty. The messy, transnational flow of electronic communications blur the "domestic" and "foreign" until it is technically implausible to monitor only the latter and never the former. Finally, implicit in this sweeping mantra is the assumption that there *must* be a needle in the haystack: that when data, such as social media postings, marriage status, and skin color, is stacked up high and correlated, they will discover reliable predictors of terrorism—that, in short, the data must hold the answers.[25]

Such data hunger was not an inevitable result of the ground facts of the September 11 attacks but rather involved a fair degree of politically expedient slippage. The 9/11 Commission reports did not directly recommend "whole haystack" collection; by connecting the dots, they meant a more traditional conception of cutting red tape and circulating existing information more effectively across intelligence agencies. One could even argue that the turn to bulk surveillance undermines the original recommendation that known suspects and known trails are covered more thoroughly. As these new systems gathered a far larger trove of suspicions, the burden of investigating them would increase exponentially. Consider, for instance, the national suspicious activity reporting initiative (NSI), created to systematically document offline surveillance, compiling and processing reports from patrolling officers. Created in direct response to the 9/11 Commission, the NSI's data covered not only actual crimes or convictions but also a much lower threshold of suspicious behavior; NSI instructions cite examples such as "unusual interest in facilities, buildings or infrastructure beyond mere casual or professional."[26] The NSI thus mirrors the indifferent recombination of electronic surveillance systems—and re-creates the latter's problems of redundancy and meaninglessness, producing tens of thousands of reports that have yielded almost no useful leads.[27] Such initiatives were contemporaneous with the expansion of electronic mass surveillance and shared a common picture of the world of new terrorism: one where the biggest possible database had to be amassed and then canvassed for signs of suspicious activity that could predict actually dangerous or convictable behavior.

Even as state surveillance grew ever more expansive and indifferent, it also took steps to maintain its secrecy, thus producing the recessive relation between the public expected to know for themselves and the systems of datafication that withdraw from their phenomenological horizon. Even after Snowden's leaks, the irony remains that most private citizens cannot ever be sure what, if any, of their communications have been subject to surveillance. This forms a stark contrast to more "traditional" types of surveillance. American police surveillance, especially regarding poorer black communities from the 1970s onward, often ensured that its targets constantly encounter the naked violence of the state: house raids, summons to court, loud patrols, pat-downs, urine tests.[28] There, surveillance is in your face (and all over the rest of your body), marking bodies and streets with its power. Data-driven surveillance reconfigures these asymmetries around the problem of visibility: human subjects become more transparent to surveillance systems that become less transparent. Not only does this imbalance create a chilling effect on the population,[29] but it also poses new challenges for the accountability of surveillance. The growth of secretive, comprehensive surveillance systems is justified by future-oriented arguments that leveraged the seemingly inexhaustible and irreducible threat of post-9/11 terrorism. This is, in fact, a rather tricky case to make: How do you claim the terrorist is more unpredictable than ever while promising that new technologies can indeed predict the terrorist? As we saw in chapter 2, this involved the recessive presentation of the proofs themselves. Surveillance is justifiable not because these particular people we have monitored ended up trying to bomb the White House but because anybody (i.e., everybody) *could* one day decide to do so, and *if* such a thing happened in the future, we *would* theoretically be able to stop them. The nonoccurrence of a posited event thus acts as an unfalsifiable proof of surveillance's necessity, while the fundamental unknowability of the terrorist requires that surveillance continue indefinitely and expand indefinitely. The (potential and/or future) terrorist thus serves to embody the epistemic contradictions around state surveillance and its data-hungry, whole-haystack approach: Can a terrorist be "known" through data? Can human intentions, beliefs, morality, be predicted through correlations of behavioral indicators?

The Lone Wolf

> He was a dude you could always just vibe with. He liked *The Walking Dead* and *Game of Thrones*. He couldn't have been sweeter. He smoked a copious amount of weed. He won a $2,500 educational scholarship. He was one of the realest dudes I've ever met. He was just superchill. He was smooth as fuck. He was not a loner. He's not anybody *like that*. I mean, he was quiet—but not in an alarming way, he was just soft-spoken. He's a Muslim, but not so religious. He was so, so normal, no accent, an all-American kid in every measurable sense of the word.
>
> He stopped listening to music. He quit drinking and smoking pot. (He started praying more, and visiting Islamic websites.) He became anti-fun. (He went to Dagestan for six months.) He grew a beard. He criticized U.S. foreign policy. "There are no values anymore," he once said. He would start failing classes. In the aftermath, we know that we never really knew him. The contents of his closely guarded psyche may never be fully understood. It's weird, they all agree. But I can't feel that my friend is a terrorist. That Jahar isn't, to me.

Such were the words of friends, acquaintances, investigative reporters, professors, and police workers—the authoritative "experts" in such a situation—about Tamerlan and Dzhokhar "Jahar" Tsarnaev.[30] They were the "Boston bombers": two Chechen-born, longtime Massachusetts-resident brothers who exploded pressure cooker bombs at the Boston marathon just two months before the first Snowden leaks. In their wake, there was a frenetic public effort to rationalize this apparent cohabitation of radical terrorism and "normal" American life. Whereas Tamerlan had already become distant and conservative after his hopes for a career in boxing were thwarted, Dzhokhar, witnesses emphasized, would watch HBO shows, smoke weed with (white, middle-class, well-adjusted) mates, enroll in extracurricular activity—as if such things should have marked him as a normal, predictable American, someone who will see the world as I do, value another human's life as I do, reject conspiracy theories as I do. (Dzhokhar subscribed to the theory of 9/11 as an inside job.) In the same way that America's presidential candidates stuff themselves with hot dogs on the campaign trail, there is a search for a basic intelligibility or affective connectivity that allows us to try to say that this person is safe, that this person is not a criminal, terrorist, murderer, molester.

Individuals such as the Tsarnaevs were so problematic because they confounded the perennial (and never quite successful) effort to cleanly separate enemies from allies, threats, and suspects from normal citizens. When the killer comes trained, equipped, and identified by a group such as Al-Qaeda, it may be easier to declare: this individual's being is entirely and, from the beginning, coterminous with violent terror. The category *is* him (and it is almost always a "him"—statistically and in the public imagination). To know Jahar, on the other hand, requires a narrative of a switch, a doubling. He must have begun as a benign, *real* American— and then somehow changed into a killer. Something must have made him trade pot for bombs. Or perhaps he had always harbored, in the schizophrenic folds of a duplicitous soul, a Jihadi mind: a double, an infection of the unknown, the other, the dangerous, in the American body politic.[31] Dzhokhar Tsarnaev brings into stark relief this connection between knowledge and uncertainty, between the narrative of unpredictable terrorism and the data hunger of new surveillance systems.

* * *

Everyone called him "Abs." He gave out Halloween candy to children and taught them how to play Ping-Pong. He invited his neighbours to barbecue.

But Khurum Shazad Butt was not the typical resident of the East London neighborhood of Barking. He dressed in the religious gown of a conservative Muslim—with a tracksuit and sneakers underneath. He turned up in a Channel 4 documentary, "The Jihadis Next Door." And now London's Metropolitan Police have identified him as one of the three

men who carried out the deadly terror attack on Saturday at London
Bridge and Borough Market.

In June 2017, Khurum Butt and two others attacked the ordinary citizens of London, and were killed in the immediate aftermath.[32] The
four years since the Boston bombings had seen numerous such attacks
across America and Western Europe, carried out by individuals as var-
ied as refugees (or, in cases, those masquerading as refugees), security
guards, and university dropouts. Yet the question remained: How could
this monster be so humane? How could a human be so monstrous?
What properties, data points, correlations, divide a normal citizen
from a dangerous one, a bad seed, a ticking time bomb? In such coverage, the sticking point was that these killers turned out to have left
hints, that there must have been opportunities to predict and prevent
them. Butt had previously featured in a televised documentary about
jihadists, and two citizens had alerted authorities about his behavior
in the months leading up to the attack.[33] The police deemed him "low
risk" but, surely, the media narrative went, more could have been done.
Such retroactive assessments enact pseudo-therapeutic debriefings for
the public, addressing the lingering question: How could this have happened? Like the public inquest following the September 11 attacks, they
normalize the narrative that if only more dots could be gathered and
connected, terrorism might be stopped, that these deaths could have
been prevented.

Foucault wrote that a rationality exerts itself most strictly, most explicitly, not among the normal subjects who have internalized its reason but with the mentally ill, the criminal, the freak—the entities that
threaten the boundaries of the possible.[34] The excluded are not those
that lie outside reason but those who are dominated by it. Where dangerous subjects have already escaped direct containment (having succeeded in carrying out their attacks), this public, collective analysis
of their psyche seeks to restore the sense that terrorism can be predicted and that a rational comprehension of the nation's population
is possible. And so, these killers—sometimes construed as insane and
psychotic beyond comprehension, sometimes closer to troubled souls
in an accidental spiral of evil—become the focal points of knowledge
production. Tsarnaevs', Butt's, and other such attacks over the 2000s

and 2010s were analyzed through a particular ideal figure: the "lone wolf" killer, "a person who acts on his or her own without orders from—or even connections to—an organisation."[35] Such a definition seems to at least partially exclude the likes of Butt, whose connections to Daesh (aka ISIS, ISIL, the Islamic State) were relatively concrete.[36] But the lone wolf is as much the formulation of an affect as an institutional classifier—a figuration of vulnerability and strangeness at the heart of the nation. Like "national security" or "terrorism," it is a flexible idea that the state uses to draw and redraw the definition of the enemy to meet its own strategic needs. Even as experts expressed concern over the lack of a clear definition,[37] what percolated publicly was the sense that the world we live in is seeded with unpredictable, sudden danger. It mattered more that "Abs" played Ping-Pong with kids before heading out to mass murder than the detail that he acted with two co-conspirators.

If the Snowden files materialized the recessive relation between the public and state surveillance, the lone wolf performs a similar function at the intersection of terrorism and the body politic. Historically, the figure of the lone wolf itself emerged from a desire to (and a fear of) remain unpredictable, to become statistically indeterminate. It originated with not security experts but with white supremacists Alexis Curtis and Tom Metzger, who popularized the term in the 1990s to project the "evolution" of their struggle from easily identifiable groups to anonymous individuals that act alone.[38] The security establishment's uptake of this label marks a qualitative shift in the imaginations of terrorist danger. As we can see with Dzhokhar Tsarnaev, the uncertainty that plagues the lone wolf concerns not simply the generic threat of terrorism or violence but also a more viral danger: a diffused vulnerability to the smallest of human units, where just a single outwardly normal youth is enough to puncture the illusion of safety through knowledge.[39] The lone wolf, then, is a ghost, a figuration of that which lies outside the epistemic systems of surveillance and security. Like the residual in statistical analysis, it expresses the indeterminate danger that remains after every effort to predict and prevent. This fear of an indiscriminately distributed potential for terrorism would feed into the idea that state surveillance, too, must become more comprehensive, more indiscriminate, and infinitely data-hungry.

This expression of increased uncertainty, once again, is met by a double movement typical of the recessive relation. Even as the lone wolf becomes the bogeyman of the indeterminable, it is made the object of intense academic, state, and popular effort to "figure it out"—to fold this alien threat back into the domain of the knowable. By the late 2000s, the lone wolf had become a prominent figure in analyses of domestic terrorism. In 2009, the US government announced a "Lone Wolf Initiative" to correlate existing data to identify lone offenders before they could act.[40] Already, a preemptive, predictive attitude had been established and hinged on the ability to "know" the individual at his most private: How can you tell what an individual is doing in the privacy of his or her own home well before committing any explicitly criminal act? As the FBI put it in 2009: "[H]ow do you get into the mind of a terrorist?"[41]

High-profile attacks in the early 2010s provoked searches for statistically significant, and thereby predictive, variables for the lone wolf. The kinds of questions these studies begin by asking already express the hope that surely, like other dangers and threats, we shall be able to render the lone wolf knowable: "What are the demographic characteristics of the lone wolf?" What are their ideologies? How are they different from other groups?[42] Here, it is not the data that "speak," but political exigency, mobilizing whatever available correlations to justify action. Such queries, however, tended to yield results that testify to just how ordinary and unpredictable lone wolves could be. Many studies sought to isolate lone wolves' "unique psychological and motivational factors,"[43] typically founded on a "faulty"—that is, nonnormative—environment. Hence, one study insists that the infamous terrorists Timothy McVeigh, Ted Kaczynski, and Eric Rudolph are united by a "repeated failure" to belong, while another speaks of a "difficult childhood" and "changes in personal behaviour."[44] One sampling showed that lone wolves in the United States have a higher-than-average chance of living alone, changing their address and/or losing their job in the twelve months preceding the attack, and having records of criminal behavior or mental health problems.[45] Yet each of these factors is shared by fewer than half of these proven lone wolves. Should the state put under watch an individual who loses their job and separates from their partner? Or someone whose problems with depression means they are struggle to work regularly? Whatever formal and general property extracted from existing lone wolves ultimately proved insufficient to identify, predict, or explain their ultimate turn to violence.

In short, the effort to know the lone wolf yielded little in the way of telltale signs, only "many weak signals"[46] too flimsy for traditional thresholds of suspicion or prediction. Jahar Tsarnaev was failing his university courses, but he was popular with his Cambridge friends. Abs Butt was on television carrying the Islamic State flag in a London park, but he had given clear no indication of imminent violence to those watching him. This lack of "clear signals" keeps the lone wolf in a state of nonspecificity, forcing the question: What if we can never know? Most lone wolf killers remain mysterious in their motivations and ideological allegiances even after the act.[47] By 2019, the Christchurch shooter could be found deliberately mocking these attempts at rationalization in the messages he left behind.[48] Across academic and policy analysis and popular media coverage (most clearly in the case of Man Haron Monis, instigator of the 2014 Sydney hostage crisis), there was an acknowledgment that the lone wolf does not boil down into a single profile, personality type, or even definition.[49] Instead of the Al-Qaeda operative or the card-carrying white supremacist, the lone wolf expresses the latent potential for *anybody to become a terrorist*—a structural paranoia that is formally analogous to the paranoia of data-driven surveillance writ large: not needles in a haystack but that every piece of hay could become a needle when you look away.

Yet the desire to know, to predict, to stabilize persisted. The lone wolves refused to yield simple and clear correlations for statistical analysis, but the political and moral pressure to capture it remained powerful. To fill this lacuna, assumptions and archetypes were integrated into a disavowed form: "the lone wolf is impossible to predict or identify, but even so . . ." In the process, the "anybody" of the potential terrorist was flattened into the Arab and Muslim.[50] Although the term itself originated with white supremacists and attacks committed by white Americans continued,[51] the formally unknown body of the terrorist in our midst was silhouetted with brown and beard. Broad geopolitical perceptions thus find their way into working approximations of a technically unpredictable group—to the point that "recent conversion to Islam" alone (rather than religious fundamentalism in

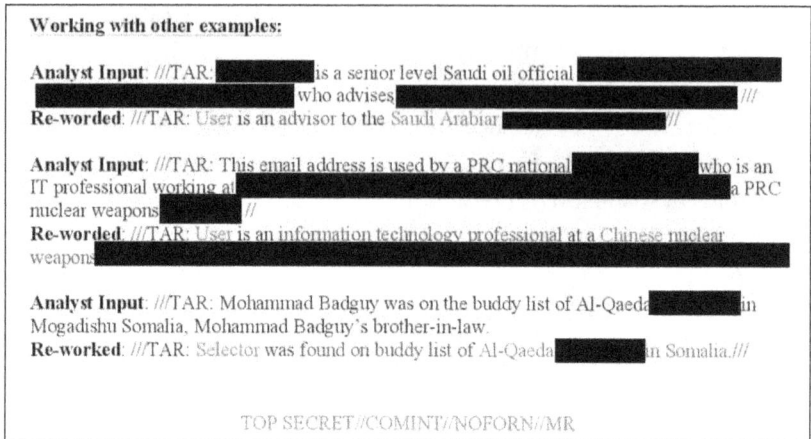

```
Working with other examples:

Analyst Input: ///TAR: ███████████ is a senior level Saudi oil official ███████████
███████████████████████ who advises █████████████████████████████████████ ///
Re-worded: ///TAR: User is an advisor to the Saudi Arabian ████████████████████ //

Analyst Input: ///TAR: This email address is used by a PRC national ████████ who is an
IT professional working at █████████████████████████████████████████████ a PRC
nuclear weapons ██████████ //
Re-worked: ///TAR: User is an information technology professional at a Chinese nuclear
weapons ███████████████████████████████████████████████████████████████

Analyst Input: ///TAR: Mohammad Badguy was on the buddy list of Al-Qaeda ████████ in
Mogadishu Somalia, Mohammad Badguy's brother-in-law.
Re-worked: ///TAR: Selector was found on buddy list of Al-Qaeda ████████ in Somalia.///

                    TOP SECRET//COMINT//NOFORN//MR
```

Figure 3.1. Examples from an internal Targeting Rationale document leaked by Snowden. Source: Document available through the nonprofit National Security Archive, "Targeting Rationale," National Security Archive, n.d., https://nsarchive2.gwu.edu/NSAEBB/NSAEBB436/docs/EBB-125.pdf.

general or specific extremist doctrines, such as Salafi jihadism) become cited as possible predictors for lone wolves.[52] In 2017, a leaked document detailed a survey used by FBI agents to determine the risk of violence posed by suspects.[53] Alongside questions about "weak indicators" (Has the subject participated in activities such as paintball or laser tag?), it instructed agents to ask questions: Has the subject undergone religious conversion recently? Has he or she "articulated a desire to conduct violent jihad or achieve martyrdom"?[54] To be sure, the threat of terrorism by self-professed Muslims of Arab descent was an immediate and acute one for post-9/11 America. The point is that this specific figure of the Muslim extremist often became conflated with, and stood in for, the allegedly unknown lone wolf. This is exemplified by Mohamed Badguy, a fictional name used in NSA slides to provide examples of the targeting process. He is joined by an equally cartoonish "Mohammed Raghead," which the NSA used as a placeholder for surveillance memo templates (figure 3.1).[55] As we will see in chapter 5, scenarios, exemplars, and simulations reveal much about what kinds of assumptions and provisional designations "stand in" for the as-yet unknown.

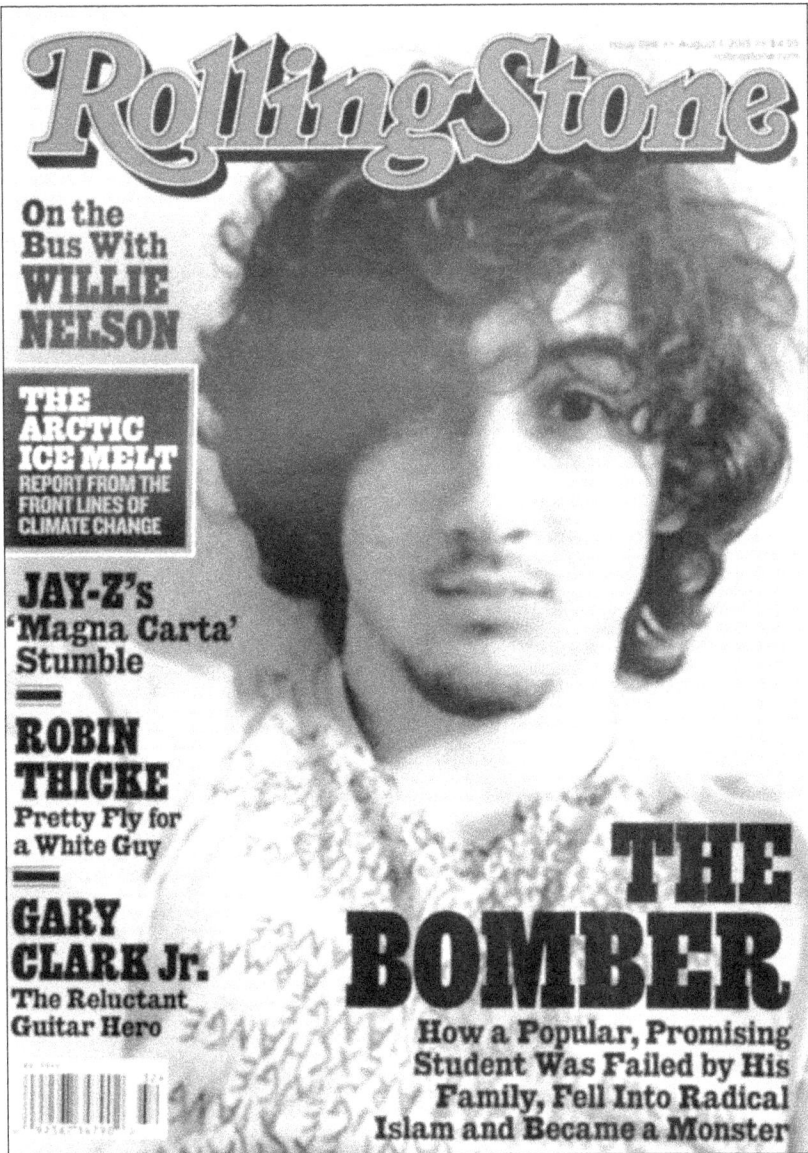

Figure 3.2. Dzhokar "Jahar" Tsarnaev on the cover of *Rolling Stone* in 2013.
Source: Janet Reitman, "Jahar's World," *Rolling Stone*, July 17, 2013, http://www.rollingstone.com/culture/news/jahars-world-20130717.

Purifying the Body Politic

Four months after the Boston bombings, Jahar Tsarnaev made the cover of *Rolling Stone*. His face received a "dreamy," "glam" treatment[56] that recalled rock stars and other acceptably transgressive youths who had been the magazine's bread and butter (figure 3.2). Predictably, the cover generated a great deal of outrage; a local councilor lamented that the killer had been "marketed as a hero, a misunderstood teen." The controversy around the humanity of Jahar reflected the fragility of the boundaries securing the normal from the deviant. Western discourse on the terrorist has frequently invoked the ancient frame of the monster, the disturbing entry of fundamentally unintelligible and strange deviance into the normal and the internal.[57] This becomes visible in the tendency to classify each new terrorist as psychologically insane and/or an inhuman embodiment of evil. Such classification often occurs very quickly, before even the basic timeline of events has been accounted for, and the activity is not only emotional but heavily moralized, with accusations of terrorist sympathies leveled at politicians or citizens unwilling to go the distance. This injunction to classify and condemn responds to the uncertainty raised by each new attack, such that the attempt to write each killer off as crazy or evil reflects the felt need to somehow define and reduce these frightening possibilities. In fact, the boundary work performed here is a straightforward extension of what occurs in more "banal" genres of reality television, crime and police shows, or the routine circulation of viral stories online about crime and stupidity.[58] Encouraging audiences to laugh at idiotically failed robberies or to condemn drug-addicted fathers leaving their infant children to die, such media perform the everyday separation between the dregs of humanity in our midst and the normal, upstanding citizens. This neat, sanitized splitting of the population in the cultural imagination allows the crisis of unknown dangers to be folded back into the existing order. What cannot be known and accounted for, can at least be cleaved away as the heart of darkness in the global social fabric.

As Foucault showed, any regime of knowledge is defined by what it considers knowable and what it renders invisible and unsayable. This is not the distinction between what statements are accorded the status of truth or falsehood but what kinds of statements are eligible to play the game in the first place. Judith Butler derives from this a matrix of inclusion/exclusion: to be a subject is not necessarily to be what dominant power relations tell you to be but to define your being in their terms.[59] Often, the price of refusing this bargain is not to become a resistant or alternative subject but to become *unintelligible.* As a label, the lone wolf describes American society's effort to take the unintelligibly Other—a being which transgresses the basic obligations of citizenship—and somehow reintegrate this unknown into grids of intelligibility, even as this very move forcefully restates the monstrous otherness of the unknown. It is a practical and strategic need, as the state tries to figure out where to devote its resources and which kinds of civil rights to suspend in what cases. At the same time, this pursuit of this unintelligible monster subjects everyone else to surveillance's regime for converting bodies into facts—a process that does not take hold equally but treats different kinds of bodies to different kinds of factmaking.[60]

This boundary work, managing the limits and directions of security practices as well as public moralizing, becomes all the clearer when we consider the changing meaning of the lone wolf itself. Having become a household word through the figuration of Arab Muslim extremists (and thus forgetting its origins with white supremacists), the lone wolf remained definitionally unstable. The question of whether a given terrorist was a lone wolf, and whether they could have truly been preempted by data-driven surveillance, was posed less as a technical question than as a heavily moralized and political one. In 2017, Stephen Paddock, a white sixty-four-year-old former accountant, shot into the crowd at a music festival from a Las Vegas hotel overlooking the venue. Paddock's use of modified automatic weapons ensured that it was the deadliest shooting in modern American history, with more than fifty deaths and five hundred injured. Yet Paddock had not expressed Daesh sympathies,[61] had no history of mental illnesses, and indeed left no evidence of a coherent motive, killing himself before the police could apprehend him. It was a violent reminder that white men, rather than young, antisocial Muslim youths, have continued to be the single largest contributing demographic to mass killings in the United States. In the wake of the attack, as politicians and the public again sought to categorize what had happened, the category of the lone wolf became a hotly contested object. Donald Trump reprised the stereotype by calling Paddock "sick and demented" in the absence of any evidence or diagnosis. State authorities chose to characterize Paddock as a lone wolf—but not a terrorist. The lone wolf thus functioned as a way to reserve judgment about the killer, effectively shielding him from the rapid separation as a monster that other shooters had been subject to.

This usage of the lone wolf was quickly criticized. Commentary across *The Washington Post* and other outlets argued that the category functions as an "out" for white killers: a relatively benign and noncommittal characterization that takes a softer touch compared to "terrorist."[62] Juxtaposed against the quick suspicion of every brown killer as "terrorist," the lone wolf now appeared symptomatic of the country's wider failure to recognize the clockwork repetition of mass killings by white perpetrators. The usage of the lone wolf as an evasive category again demonstrates how such acts of identification are based less on the realist emergence of novel threats and more on the shifting political needs for the right kind of silhouette. The state authorities' reluctance to use the label terrorist on Paddock, despite many definitions at state and federal levels employing a broad definition of the term with no religious or ideological qualifier,[63] is telling. Having enacted a recessive relation with the uncertainty of "new" terrorist threats, the figure of the lone wolf was now used to occlude the truly diffuse nature of this threat—instead isolating it into the mad, the evil, the nonwhite, the Other.

And so, from the white supremacists of the 1990s through the Mohammed Badguys to the whiteness of Paddock, we come full circle. The lone wolf is clearly neither white nor brown in essence; it bends to political will and accumulated prejudice. The efforts to datafy, categorize, correlate, predict, entail a promise to leave behind biases and politicized assumptions. But this very promise of epistemic purity, this "sterility" of data-driven knowledge, is what allows bias and politics to smuggle their way back in. In the war on terror, the lone wolf was an instrument for dividing bodies into safe and dangerous, humane and monstrous, citizen and enemy—an act of division that, of course, subjects the entire population to measurement and observation. It betrays a desire for purity, for sterilization, for separation, which Latour would tell us is baked into modern systems of knowledge: to cordon off the unknown and declare the inside a clean space where the bad guys are kept out. The act of distinguishing Paddock as a disturbed lone wolf from the brown terrorist serves as a way to quarantine this white killer as a deviant case, a not fully human case: he is just a crazy wacko. This rhetoric preserves the essential distinction between American Us and Muslim extremist Others. The modern idea of security is, at its core, a fantasy about certainty through sterilization, one that equates purity with safety: put a hermetic seal on the borders, ensure everybody on board is accounted for, and quarantine every unpredictable element. Here, we find one important set of political stakes to technologies of datafication. This is not to say that Bayesian methods or statistical prediction inevitably result in an art of government that separates, quarantines, excludes, and eliminates. But databases offer ways to cut and cordon populations above and beyond the muddy ways in which fleshly bodies circulate, and predictive recommendations offer ways to categorize and determine individuals well in excess of other forms of evidence-based judgment.

The nexus of data-hungry surveillance systems and the monstrous figure of the lone wolf operationalize uncertainty—not by resolving it into calculable certainties but by legitimating new public rituals of purification and rationalization. Enduring social imaginaries around terrorism and deviance, America and race continue to feed into the speculative folds around surveillance and its data. The technically unknown outline of the lone-wolf killer is thus silhouetted into the Arab Muslim (whether frighteningly sober or dangerously deranged) or the mad white man as exception. Uncertainty is not a void but a space for filling in with convenient truths.

Such operationalization of uncertainty serves immediate practical and strategic functions. It allows for the justification of expansive surveillance, the creeping extension of everyday racism, or a rhetorical and affective repair of the sense of normality that allows us to keep going out onto the streets the day after an attack. In chapter 1, I linked the technological fantasy of data's purity to its bid to become society's new groundless ground: the unprovable, collective belief that a stable and objective reality exists out there and that data can mine it, extract it, and bring back home a slice of that knowledge. The fabrications surrounding the lone wolf manifest this fantasy in concrete ways. Contrary to its initial appearance as a figuration of the unknown, the lone wolf soon became useful as grounding for state surveillance and the war on terror. The journeys into the deep interior of the killer's psyche involve a search for the kind of "raw" truth that can authenticate our theories, rationalize the unintelligible, and maintain the collective belief that better knowledge is possible. Rationality and knowability are protected not by an impossible eradication of all that lies outside it but by designating the latter to its rightful place. In the high-stakes context of lone wolf terrorism, the pursuit of datafication is driven by the idea that there *must be* some correlation, some data point, that can allow us to decode the hidden logic that turns "troubled teens" into suicide bombers. This necessity of a grounding leads surveillance more generally on a constant search for whatever is most amenable to observation, collection, quantification, and correlation. Yet this data must also remain authentic, "raw," maximally protected from the vagaries of human interpretation. Hence, the investment into biometrics and other traces of the body as the next frontier in state surveillance—and, as we shall see in the next chapter, the concurrent boom in the business of self-surveillance.

The Body That Speaks

The pursuit of the lone wolf began with the projection of a radical uncertainty and then a series of attempts to operationalize that uncertainty through ever-expanding, ever-hungry surveillance systems. These systems depend on the belief that the human body will and must speak: that its movements, its actions, and its demographic properties will eventually yield calculable patterns and predictive certainty. Investigations into the psychic life of the Other contribute to this projection of the natural or objective body, insofar as it seeks an empirical source that speaks the truth and that can be forced to speak the truth, independent of the subject's intentions.

The body, of course, does not always speak—or, at least, not in the way we want it to. Predictions fail, and terrorist attacks succeed. What then? As we saw with the likes of Dzhokhar Tsarnaev, a protracted public debriefing ensues. They emphasize the could-have-beens, the missed opportunities: we just needed to gather more data, connect more dots. Indeed, such debriefings function as paradigm repair, restoring the general narrative that surveillance is necessary and datafication will find the truth. Chapter 5 examines the results of this doubling down: a rationality of "zero tolerance," which again fields the fantasy that the law and the state might impose a pristine, error-free grid of knowledge on human bodies and intentions.

In this pursuit of predictive material, the human individual becomes Deleuze's dividuals: people cut up into a wide array of data points, split and fed into algorithms.[64] The recessive ambivalence around the lone wolf corresponds to this delegitimization of human-infected truths. We cannot know the lone wolf, for human subjects cannot be trusted to be sufficiently honest, sufficiently rational, and sufficiently consistent to provide the early warning signs. Meanwhile, an alternative production line for reliable facts emerges: the machine, with its boundless data amenable to infinite recombination, harvesting from the body what the subject is unwilling or unable to tell.

At the same time, these dividual points constantly refer back to the presumed unity of the body. There is a constant effort to take these dividualized data points and use them to reassemble, a coherent body, a coherent subject. This paradox comes to the fore in the case of self-surveillance, whose promise of knowledge for self-improvement hinges on this undercutting of the conscious subject. In contrast to the Big Brother overtones of government surveillance systems, self-surveillance is often promoted as a democratization of big data technologies. The search for knowledge turns from the "out there" to the "in here," from the secrets of *arcana imperii* to the mystery of one's own body. Here, we find the same anxieties around uncertainty, complexity, and visibility—but now woven into the problem of "knowing oneself" in our private and domestic lives. Even as the public is called on to fulfill its duty to know for itself and the individual is enjoined to know thyself, we find that the good liberal subject is increasingly divested of the authority it once supposedly commanded.

4

Data's Intimacy

Walking, sleeping, talking: it's the stuff of everyday life.
Add sensors that track all of it, and suddenly everyday life
becomes an opportunity for knowledge.
—Nick Wingfield, "Gauging the Natural, and Digital,
Rhythms of Life"

You sleep.[1] A thin, rectangular strip, slipped unobtrusively under the
bedsheet, senses your arrival, your movement, your resting heart rate,
your respiration cycle, and more besides. The smart sensors on the
strip collect the data and transmit it wirelessly to the cloud for analysis.[2] When you wake, your consciousness is greeted by a numerical sleep
score on the smartphone app screen: a simple distillation of the many
data points, and their estimated relationship to sleep quality, into a score
out of a hundred. The information collected by such machines remain
fairly rudimentary and often woefully ignorant of context. Yet even as
the devices remain imperfect, their deployment already embeds a certain
communicative network into the rhythm of everyday life. The machine
delves deep into the body, leveraging the latter's constant, nonconscious
discharge of material traces. In doing so, these devices promise to measure what the human cognition and intuition cannot. Only afterward,
bleary-eyed, does the conscious subject enter the picture.

* * *

June 2015, San Francisco. Ariel Garten, the same entrepreneur who had in
the introduction spoken of humanizing technology, was now explaining
her flagship product to an audience of Quantified Selfers (QSers).[3] A thin
headband would pick up electrical activity in the brain by adapting electroencephalography technology well known in neuroscientific research.
The hardware is then paired with a smartphone app, where signs of distraction and stress will trigger the sound of strong winds, and a calmer
brain will soothe them. Garten explained that the meditation augmenting

device "gives you an experience of yourself. It's not a quantification, it's not a number, it's not a data point . . . you're really hearing yourself for the first time." This empowering rediscovery of the self depends on machines that speak to my body in ways that I cannot—and, Garten hoped, machines that would eventually go on to optimize what it sees. This intervening, nudging, conditioning relation seeks to establish the technology not as an instrument but as a part of the subject's prereflective equipment for sensing their own bodies, the "feeling of their own feelings."[4]

Today, machines observe, record, *sense* the world—not just for us but sometimes instead of us (in our stead) and even indifferently to us humans as well. If state surveillance sought direct access to the body that speaks as a way to outwit and preempt the malicious intentions of would-be terrorists, the same principles for extracting the truth from the body are also unfolding at the levels of consumer activity and lifestyle. Such tracking serves a plurality of interests. Scholars have explored how individuals might engage with the tracking on their own terms, actively negotiating with the commercial and technological affordances in a practice of "soft resistance."[5] For individual users, the technology provides a platform for long-standing desires for archival and communication—to find new ways to remember myself, to tell stories about myself.[6] At the same time, the rapid growth of the tracking industry has hastened its integration into the data market writ large. Where specific forms of human behavior have long been regimented into formal scenarios amenable to surveillance and datafication—the testing of schoolchildren, the scoring of credit, the upkeep of medical records—everyday life presented many little cracks through which usable, profitable data could be lost. Wearables, home monitors, and smartphone apps thus seek a persistent, fine-grained capture of the personal and the mundane, persuading individuals that whatever knowledge they had formed about their own everyday health, happiness, or productivity is better off superseded by the analysis of the smart machine. This chapter argues that self-surveillance renders the personal machine-readable, enacting the duality of fabrications: the pursuit of self-knowledge reshapes self-knowledge in data's own image, in terms of not only the resulting analysis but the machines, methods, and dispositions that are required to become a persistent part of everyday life to produce that knowledge.

In seeking to advise, correct, and even overrule individuals about their own self-knowledge, technologies of self-surveillance introduce their own tendencies around what counts as data about the self and what counts as an objective processing of that data. On one hand, self-surveillance is consistently justified through the fantasy of honeymoon objectivity, in which the veridical authority of technoscience is juxtaposed to seemingly inherent irrationalities of the human subject. But what does it mean to define *rational* as something that is inhuman? In *How Reason Lost Its Mind*, Paul Erickson et al. argue that this "idea that machines might reason better than human minds was alien to Enlightenment thinkers."[7] If the human exercise of reason was an irreplaceable foundation for classical liberal thought, the Cold War saw machinic rationality, composed of formal, algorithmic rules for processing information, overtake the former in epistemic priority. Of course, a simple linear narrative in which human reasoning increasingly loses out to machines occludes the qualitatively different styles of reasoning introduced by technologies of datafication. The smart machines in this chapter advance a specific method of fabrication in which trivial traces left behind by the body and its most mundane activities are revalorized as the secret clues to one's health, happiness, and productivity. It extends and modifies a tradition of what Carlo Ginzburg called conjectural knowledge: "an interpretive method based on taking marginal and irrelevant details as revealing clues."[8] Ginzburg cites the example of the "Morelli method" in the nineteenth century: a painting's provenance would be identified not through visibly meaningful elements but through unreflective and trivial patterns, such as how the ear tends to be drawn. In this method, or Freudian psychoanalysis directly influenced by it, the clues involve a certain indifference to the foregrounded relation of meaning-making, to the conscious intentions of the creator or the experience of the patient.

Decisions of political and moral import are embedded in such methodological habits. It entails selecting what kinds of data might serve as clues, and what kind of rationalization—and by whom—might properly interpret those clues. A similar selectivity is also baked into data analytics, most clearly with artificial neural networks.[9] At one level, the machine might be "left alone" to discover what correlations in the data might be relevant for delivering the desired result, such as the classification of faces by race and gender. Underlying this seemingly autonomous function, however, are the ultimately human, ultimately groundless decisions about what kind of training data to feed the network, what kind of initial classifiers to be used, and even the basic assumption that, for instance, homosexuality can be and should be predicted on the basis of facial images. With self-surveillance, it is the traces most amenable for data-driven prediction, and the traces most cheaply accessible to existing sensors, that become privileged as the objective keys to the holy grail of self-knowledge. Through this contingent mixture of technical affordances, commercial imperatives, and convenient assumptions, self-surveillance begins to perpetuate its own style of reasoning about what counts as a clue to the truth of the self, how those clues should be analyzed, and why machinic sensibility is to be trusted over human experience.

These emergent styles of reasoning were justified—and more than justified, made attractive—through a technological fantasy of data's intimacy: the idea that machines will know us better than we know ourselves and that through these friendly devices, we will also achieve better self-knowledge. It is a kind of "knowing" that interweaves modernity's epistemic virtues of accuracy and objectivity with existing cultural sensibilities around individual empowerment, spirituality, and the care of the self in neoliberal times. In contrast to state surveillance and the war on terror, data's intimacy thus promises to bring datafication *close* to us and to reverse the worries that individuals will be left behind by recessive technologies. But once brought close, these smart machines have a way of making the personal a good deal more complicated. As such networks reconfigure the production and circulation of "personal" data, they also redistribute the actors and authority involved in the production of "self"-knowledge. Who (and what) produces data about the individual? How is it fabricated into the status of truth, as knowledge? Precisely because these smart machines stick so close to us and promise "insights" that upturn our own ideas about ourselves, my epistemic relation to "myself" becomes rerouted and externalized—betraying the old fiction of im-mediacy embedded in the word *myself*.

The expanded frontiers of datafication, carved out by self-surveillance in the name of intimacy and personal empowerment, inevitably serve as fertile ground for data extraction and recombination. This chapter traces how even as self-surveillance was popularized through the allure of personal choice and control, the mechanisms of data collection and fabrication created this way are increasingly being leveraged for populational surveillance and control by state and corporate entities. A fitness-tracking wristband, originally developed for individuals to assemble new clues about their bodily activity toward personal improvement, now supplies insurance companies with clues for the distribution of monetary risk across the client population. At the same time, the state seizes on this data as a new pipeline for juridico-legal truth, producing situations in which my own smart machine might testify for or against me in court. The result is the kind of subject, and the kind of truths, most compatible with the data-driven society's systems of labor and value production.

The World of Things

The sleep tracker (Beddit) and the meditation augmentor (Muse) were both in San Francisco that day, having turned up for the 2015 Quantified Self conference. The Quantified Self (QS), a community of self-tracking experimenters, received the name through two *Wired* veterans in 2007; by 2015, it had spread to dozens of cities globally, with regular conferences across Europe and America. As a hub for connoisseurs, enthusiasts, and visionaries, QS constituted an overlapping but distinct space from the industry or market writ large, within which a certain subcultural traffic of ideas could occur. Such spaces helped crystallize a narrative, connecting individual technologies, such as Beddit and Muse, toward a futuristic vision of an individualistic, objectively known, constantly improving, data-driven human. As Kevin Kelly, QS's co-founder and a longtime evangelist for computing technologies, put it,[10]

> [t]he central question of the coming century is Who Are We? What is a human? . . . Many seek this self-knowledge and we embrace all paths to it. However the particular untrodden path we have chosen to explore here is a rational one: Unless something can be measured, it cannot be improved.

By the early 2010s, this call to track the self was finally—after decades of premature excitement—achieving widespread popularity.[11] Market estimates pegged the sale of "wearables" (from fitness bands to smart-watches) at 15 billion USD in 2015,[12] with more than twenty million units shipped worldwide in the third quarter alone.[13] This generation of self-surveillance involved numerous overlapping kinds of technological inventions,[14] connected by their practical orientation towards the data-driven monitoring of human bodies in their most quotidian activities for improved predictivity. Some attached themselves prosthetically to the body in movement, extending on decades of experimentation in wearable computing devices.[15] Others were embedded in the domestic environment, marketed as "smart" appliances that could monitor babies or reduce energy waste. In most cases, these devices depended on two recent technological affordances. Sensor technology had improved sufficiently[16] to allow affordable and miniaturized products—which could, by the late 2000s, rely on cloud computing and wireless networks to establish frequent, around-the-clock machinic communication as a norm (at least in many parts of North America and Europe). This boom in self-surveillance was itself knitted into an even broader narrative of the Internet of Things (IoT), itself a reprisal of ubiquitous computing, ambient intelligence, and other theories of smart machine networks that predated the World Wide Web.[17] IoT was hailed as the tech trend "that will define 2014," an innovation predicted to hit the big time in 2013, a tech "about to go macro," or, even, one of the decade's defining innovations that will have "invented the future."[18] Never mind that various connected appliances, such as "smart" refrigerators, had been launched and relaunched over the previous two decades;[19] like self-driving cars or remote work, IoT and self-surveillance weathered generations of disappointments with yet another reboot.[20] These associations with IoT and big data only reinforced the contemporary perception that self-surveillance was at the vanguard of the latest technological revolution. Journalistic prognostications, industry investment, well-connected visionaries (or, as Fred Turner calls them, network entrepreneurs[21]), early enthusiasts, and media-friendly prototypes combined to illuminate a delightful and empowering proximate future—a future of better self-knowledge.

As these devices began to achieve popularity beyond the Bay Area or "techie" communities, the public presentation of self-surveillance also grew diffused and multifaceted. QS was just one part of this landscape—

and at any rate, the community itself represented no singular set of subjects or attitudes. Gary Wolf and Kevin Kelly, QS's co-founders, explicitly advocated a decentralized approach to community building and sought to provide maximally loose guidelines that would knit international practitioners into a textual community. The relative openness of QS gatherings attracted not only amateurs showing off their eclectic garage experiments but also journalists, academics, and "quantrepreneurs"[22] that might canvass the community for their upcoming commercial product. Meanwhile, the fast-growing sales of mass-produced tracking products during the early 2010s produced a large, if still nascent, industry—one that interpellated its consumers as individuating themselves through mass consumption. If the widely celebrated turn to big data involved projects of indiscriminate surveillance and totalizing archives among government agencies and of exhaustive consumer profiling, personalization, and predictive management amongst commercial platforms, then a similar set of technologies were beginning to spark a distributed landscape of connoisseurs, self-experimenters, start-ups, academic researchers, and industry heavyweights converging on a certain "personal" application.

In this emergent period, self-surveillance was part big business, part big dreams: a mix of imperfect, often prototypical devices and ebullient futurism. The latter should not be mistaken as mere rhetoric, a hyperbolic bubble frothing over a more modest reality. The production and presentation of these visions were meaningful historical actions that sought to organize how the public understood these new emerging regimes of knowledge. The question here is not to sift "real" technological innovations from mere hype or to evaluate the success or failure of this or that technology but to assess how the actual and ideal intersected to produce new epistemic relations among big data, smart machines, and individual human subjects. In his final works, Foucault sought to undertake what he called alethurgy. "Etymologically, alethurgy would be the production of truth, the act by which truth is manifested": not "What truth?" but which actors and forms accrue the status of producing truth.[23] In the case of self-surveillance, the question is how society organizes the self's ability to speak its truth, to make itself intelligible. The conditions under which individuals are encouraged to know themselves, and the technological design that configures their ability to datafy themselves, structure the ways in which we make ourselves intelligible to ourselves in the first place.[24]

Data's Intimacy

These fantasies of relentless measurement for comprehensive self-knowledge communicated a certain sense of intimacy. Where the word's recent usage in theories of media have often focused on connotations of human togetherness and affective transmission,[25] its roots in *intimus*, the inmost parts of the self, tells us that this proximity can apply elsewhere. Our understanding of intimacy as love, friendship, or other forms of human togetherness remind us that even as we hope intimacy makes us whole and puts us in touch with ourselves, it can also be a disruptive and fraught process that opens us to vulnerability and otherness. Intimacy is not just about getting "in touch" with our inner selves; it also turns us inside out. One important sense in which self-surveillance produces intimacy is that it establishes a new sense of proximity or access between the machine and the body, the machine and the body's truth—which relation seemingly authorizes the truth value of machine-extracted data. At a thematic level, we may subdivide this intimacy into a tripartite fantasy: (1) of a truly *personal* self-knowledge, superseding the population as a unit of analysis; (2) of technology not as discrete tools but a persistent and ubiquitous *background*; and (3) the *emancipation* of ordinary folk from their subjection to top-down surveillance, turning them into independent entrepreneurs of their own data. Together, these narratives promote a "healthy" skepticism of human intuition and experience and a corresponding faith in machinic sensibility.

1. Small Data

The advent of self-surveillance was positioned as an individualizing upgrade to the population as a unit of analysis and knowledge. The latter's development over the eighteenth century had constituted an effort to "know" a human multitude that was threatening to grow beyond traditional means of approximation, especially lived intuition.[26] Populational data allowed the individual to compare themselves to their proper category or even to approximate one's individual value when it could not be directly measured. Many QSers framed their own motivation to self-track as arising from frustrations with this averaging calculus. If people seem to respond to caffeine or cardio exercises in different ways, how can I figure out what "works" for me? The idea was that self-surveillance could answer, in ways that traditional, limited-sample populations could not (or could only roughly) predict how *I* personally would fare.[27] QSers argued that whereas older life-hacking techniques, such as self-help books, might insist on a one-size-fits-all solution applied to a generalized cohort, self-surveillance would help you discover what protein shakes or fish oil is really doing in your specific, individual case.[28]

The recurring insistence on the "unique data point" marks a particularly self-oriented (or, in the eyes of some commentators, "narcissistic")[29] mode of knowledge. Two centuries ago, Alphonse Quetelet had devised the average man (*l'homme moyen*) as an ideal fiction to understand each individual by; it was as if a "new kind of object" had been brought into the world, as real as a tree or the planet.[30] Self-surveillance now sought to overcome this specter. "n = 1" was a key phrase within QS and the wider industry discourse—as both an expression of a technical distinction from populational sampling and a slogan for the technology's beneficent attention to individuals.[31] Of course, things are rarely so clean-cut into binaries; with big data, "personalisation is the upshot of mass production" and vice versa.[32] I will later show how the statistical norm and other figurations of populational thinking loom large over each piece of "small data," indexing the latter in terms of the former—both through subjects who feel they need it to give meaning to their self-knowledge and through wider corporate systems of data capture and exchange.

"n = 1" illustrates the ideal of a new proximity between sensing machines and the individual body, recalling Hansen's twenty-first-century media. Smart machines seek to *bypass* the unaware, error-prone, uncooperative, and otherwise recalcitrant subject to get straight to the (allegedly) objective realm of bodily data.[33] This forms a stark contrast to techniques of confession and avowal that had often (although not always) characterized the search for individuals' truth in the West. Foucault gives us a remarkable story in this regard: François Leuret, the nineteenth-century psychiatrist, and his singular method for treating delusions. A patient stands under a shower. Leuret insists: there is nothing true in your delusional claims about reality. The patient: I know what I saw and heard. Leuret: You will receive a shower until you avow that everything you have said is pure madness.[34] The shower is ice-cold. The patient: I avow, but only out of compulsion. Another shower. I avow . . . All this, Foucault says, has little to do with *persuading* the patient and everything to do with leveraging the subject's veridical authority toward the official proclamation of their madness. Elsewhere, he notes that the physician, in general, is "someone who listens . . . not because he takes them seriously . . . [but to] cut through the speech of the other and reach the silent truth of the body,"[35] a methodology that psychoanalysis would adopt on its own terms. This is the game self-surveillance seeks to transform. In the proximate future of self-surveillance, there is no need for the subject to sign on the dotted line, at gunpoint or otherwise; the truth is always already communicated from body to machine before the subject has said a word.

2. Background

The turn to "n = 1" involves a certain precision, a narrowing onto the individual—but a precision that is enabled and accompanied by scalar expansion. In promising to administer truly individual analyses, self-surveillance zooms in on the idiosyncrasies of the singular body. Yet if this rhetoric of personalization evokes a sense of relational intimacy, we should remember that personalization in the age of new media—of targeted advertisements, algorithmically curated news content for online users—was largely achieved by and for expansive systems that integrate individual actions and tendencies into populational metrics. Small data is not at all opposed to the big. Self-surveillance's personalized knowledge has as its essential technical requirement a deeply and widely embedded system of persistent measurement, whose invisible ubiquity may now be presented as a positive selling point to the consumer. The corollary is that as self-surveillance practices multiply across different devices and contexts, they also start to disappear from individuals' experiences of embodied living.

Self-surveillance technologies thus aspire to, and increasingly realize, an environmental and atmospheric form of everyday presence: a background. They seek to become "part of the furniture," rather than stand-

Figure 4.1. The Mother device would collect data from small "motion cookies" attached to household objects. The product, advertised as a "universal monitoring solution," was briefly available between 2015 and 2018.
Source: "Mother." sen.se. Accessed March 29, 2016. https://sen.se/mother/.

ing out as discrete and actively used tools, spatially bound archives, or specific and purposeful queries. By the early 2010s, devices were beginning to accompany users to the bed and the bathroom, in their walks up the stairs as well as runs in the park, in their phones, and, literally, under their skin. Previously, it was more likely that measurements would be confined to specific and comparatively stable classes of objects and situations; the bathroom weight scale, the diary or journal, the doctor's office, and even the relative discreetness of the desktop computer. The shift toward ubiquitous sensors and prosthetic devices entails a qualitatively distinct form of surveillance.

Consider the ominously named "Mother"—whose branding appears a conscious play on the trope of Big Brother. Mother involved attaching a pack of small, nondescript "motion cookies" to domestic objects, such as toothbrushes and pillboxes (figure 4.1). The cookies' motion, temperature, and proximity sensors would continuously monitor whether the keys have been picked up or the front door has been opened. Although each given implementation is rather nonspectacular, such tools point toward a domestic environment where tracking passes from a specific action to a general fact. As one industry insider put it, a "planet with a nervous system."[36] Self-surveillance thus begins to implement, in concrete terms, another small part of a fantasy that has existed since at least the 1980s: of computer technologies made "invisible," melted fully into human being-in-the-world.[37]

> The most profound technologies are those that disappear. They weave themselves into the fabric of everyday life until they are indistinguishable from it . . . Hundreds of computers in a room could seem intimidating at first, just as hundreds of volts coursing through wires in the walls did at one time. But like the wires in the walls, these hundreds of computers will come to be invisible to common awareness.

Such descriptions emphasize a certain convenience—a seamlessness of use that Erich Hörl describes as a "sinking back into the inconspicuousness of the 'ready-to-hand.'"[38] Not quite, or not simply so. In Heidegger's classic formulation, *vorhandenheit*, presence-at-hand, describes situations where tools protrude into conscious human experience; in *zuhandenheit*, ready-to-hand, they withdraw into the background of use. In the case of self-surveillance, the primary distinction here is between objects that present themselves as sets of *instrumental* functions, to be activated by the user-subject, and an environment that prefigures, anticipates, and operates independently of the human individual. Furthermore, smart objects build robust communicative channels between themselves while often bypassing the human subject. In "anticipating" and "responding to" signals, as the digital assistant Siri does to a human voice, these devices effectively "modulate the intelligibility of space and time" for humans and non-humans around them.[39]

These environments extend the well-documented internet-age tendency toward a phenomenology of distraction and abundance.[40] By building a complexity of automated observations and communications across various sensors, the human user is positioned not as a centralized controller but a responsive actor who is alerted, interrupted and otherwise "lead on" by the smart environment. One entrepreneur put it in terms of a "practice of noticing"[41]—a machinic habituation of human attention that would cultivate the desired (in this case, productively focused) consciousness to begin with. From a business standpoint, backgrounding was also important for cultivating a long-term market beyond loyalty to a single product or style. Many QSers and lay users initially turned to self-surveillance to resolve specific problems with their health or productivity; a widely aired note of caution was that for every tracking device that sold millions, hundreds of thousands were being quietly shelved and forgotten once the honeymoon was over. Fitbit, the exercise tracking wristband at the forefront of the push toward a broader market, was at one point nicknamed "Quitbit" for its low rate of user retention.[42] These weaknesses pushed industry actors to search for new ways to embed self-surveillance into an open-ended, indefinite relationship.[43] Even as self-knowledge became more comprehensive and ubiquitous than ever, it also began to recede into the background and out of subjects' conscious engagement. To engage a machinic reading of ourselves is not so much to turn on and tune in but to find oneself in an always already ongoing swarm of active objects—harvesting us and communicating with our bodies (and each other) in cables below our feet, radio wavelengths beyond our senses, and frequencies beyond our temporal range.

3. Democratization

Such backgrounding supports increasingly autonomous technical systems, whose production of data-driven knowledge undercuts and bypasses the human subject to whom this knowledge nominally "belongs." Packaged in terms of personalization and individuality, however, tracking technologies were idealized as an empowering democratization of data-driven surveillance. Prominent spokespersons such as Gary Wolf positioned early self-trackers as analogous to the countercultural influence upon computing between the 1960s and 1990s. Just as those "hippies" had taken a military-industrial technology and helped produce a culture of personal computers and "digital utopianism," self-trackers would act as vanguards for turning the tide of big data toward empowerment and democratization: personal computing "all the way in" to the self.[44] Accordingly, co-founders Wolf and Kelly describe the origin of QS as a spontaneous and grassroots process—the rhetoric being that they merely "noticed" a new and clever practice already emerging around them.[45] In 2011, at the very first QS conference, Wolf introduced the community in these very terms:[46]

We saw a parallel to the way computers, originally developed to serve military and corporate requirements, became a tool of communication. Could something similar happen with personal data? We hoped so.

In this vein, he argued that self-surveillance could become a way to take "back" our data that states and corporations have been using against ordinary folk:[47]

> [W]hy shouldn't you have access to the traces of your own behavior that you leave behind and that others collect? . . . "It's more of a cultural shift," she says. "It's about creating a culture where we own this data. This data is ours."

Such sentiments were shared across a broad coalition of actors, from QSers to journalists and lay users. One entrepreneur, whose company provides microbiome analysis services for individuals, argued that whereas the individual was left at the "periphery" of the traditional health process, crowded out by experts, he could now "test things on myself, I know what is happening, I am not the body that the scientific and medical establishment acts upon."[48] Laurie Frick, an engineer-turned-artist, argued that tracking is now a given fact of social life: "I think people are at a point where they are sick of worrying about who is or isn't tracking their data . . . I say, run toward the data. Take your data back and turn it into something meaningful"[49]—in her case, personal data diagrammed into art. It is *your* data, many self-trackers insisted, so surely you should get as much use out of it as the others do.

This rhetoric of empowering personalization, however, occludes self-surveillance's participation in the broader data market. As the industry grew over the 2010s, tracking increasingly came to operate through a suite of mass-produced devices (many of them with only partial and ad hoc provisions for data privacy) through which its producers can develop large-scale populational databases of personal metrics. During the 2000s, the "Web 2.0" era had produced a Faustian bargain whereby the individuals' ability to socially connect to each other was the very means by which their data became connectible and monetizable for these corporations.[50] In the same way, self-surveillance's ability to "know" the subject is producing its own grid of legibility for a broader set of actors—from employers interested in optimizing its human resources to insurance companies that are already offering discounts in exchange for access to subscribers' Fitbit data.[51] The project of knowing more about and optimizing the self enrolls many different commercial, business, and government interests.

Even at the level of individual usage, the very promise of individual empowerment through self-knowledge valorizes a machinic environment that surrounds, bypasses, and structures a priori the very conditions of the subject's experience of self-knowledge and self-improvement. This yields a common refrain in tracking discourse: *you cannot lie to yourself anymore:*[52]

> "For a certain type of person," says Wolf, the Quantified Self founder, "data is the most important thing you can trust. Certain people think a feeling of inner certainty is misleading." . . . Computers don't lie. People lie.

What seems an entirely banal idea—of course people lie to themselves—propagates a certain framework for thinking. Insofar as human self-understanding is characterized as *lies*, human subjectivity and its tendencies become the object of suspicion and management. Such problematization is consistent with the popularity of behavioral and cognitive scientific explanations for human idiosyncrasies today. Best-selling books such as Richard Thaler and Cass Sunstein's *Nudge*[53] have perfected a narrative genre: here is how humans think they behave, here is how they *really* behave, and here is the counterintuitive scientific explanation why—for everything from how plate sizes determine how much we eat to how a majority of people believe they are smarter than the populational average. What you thought you do by choice, the story goes, your body determines through biological and physiological factors.[54] Of course, the formulation of self-knowledge as best served unbiased and emotionless itself implies a certain emotional profile for the imperfect human subject: one who is lamentably prone to self-flattery, denial, laziness, and, all in all, an inevitably human cowardice in the face of reality.

Here emerges a distinction between the self that lies and the self whose truth is disclosed by machines, between the thinking, feeling I and the physiological, behavioral, neural self that is coterminous with the body that speaks. For the Snowden affair and the world "out there," the problem was the phenomenological gap between the subject's personal lifeworld and the expansive horizons stretching out beyond its limits. Here, an analogous gap appears between the limited senses of the human subject and the deeper truths of the human body. We may knowingly tell fibs to ourselves about eating too much and running too little; we may not even notice that the eight-hour workday just included three hours of web surfing. But, the idea goes, the data will not filter out your momentary indiscretions, your corner cutting, and it certainly will not parlay with your pitiful excuses. Human memory, consciousness, and reason so often are a cursed fog on clear sight; data, unforgiving and unyielding, will scatter the confusion.

Small data, background, democratization: across each of these facets, the fantasy of data's intimacy exhibits a certain ambivalence. On one hand, there is the ideation of technology as an invisible servant, a neutral set of scalable techniques that individuals can control and use however they like. The more intimate technologies of datafication become, the more seamless and smoother they are hoped to become as well. Self-trackers were often accused, especially early on in the decade, as hopeless narcissists. Yet the true narcissism here is not some vain self-adoration, of Narcissus fawning over his own image, but the essentially self-centered belief that the technological apparatus surrounding me serves at my pleasure and under my control.

At the same time, these promises of democratization and personalization pave the way for a more comprehensive system of populational capture. By comprehensive, I do not mean that the technology proceeds to gobble up more parts of empirical reality until it has achieved some totality of the data—although that, of course, is the implicit fantasy at hand. Rather, self-surveillance colonizes the kinds of activities and spaces that were previously left to degrees of underdetermination:[55] the pockets and margins where we could leave the time card unpunched, our stories not cross-examined, the calories uncounted. In those margins lay another sense of the "personal"—the personal details, the private moments, that were deemed unreasonable to disclose to such harsh calculation. If the injunction to transparency had externalized the work of government onto the public, the valorization of self-surveillance as empowerment similarly introduces new labors, new sites of judgment,

for the individual. Here lies the catch in the promise of "democratization": now that we have given you the tools, you must take charge of your
own observation, analysis, and optimization—for your own betterment!

Machinic Sensibility

> SLEEP TRACKING APP: I see you're not violently throwing yourself
> around your bed, you must be in a deep sleep. Sweet dreams, buddy!
> ME: I'm actually still awake.
> SLEEP TRACKING APP: But you're lying still . . .
> ME: Because I'm trying to get to sleep.
> SLEEP TRACKING APP: You mean you ARE asleep.
> ME: I really don't.
> SLEEP TRACKING APP: You're going to have to trust me, I do this
> professionally and I know sleep when I see it, and I'm pretty sure
> you're asleep right now.
> ME: I couldn't be more awake.
> SLEEP TRACKING APP: This is all a dream . . .

This satirical piece featured in the QS website's "What We're Reading"
section,[56] a nod to the fallibility of surveillance technology and the
emerging tensions in the human–machine relation. When machines
cut directly to the body, the thinking subject is left, paradoxically, in
a peripheral position to their "own" self-knowledge. Self-surveillance
presents itself as an instrument of convenience, efficiency, optimization, transparency—that is, a neutral reduction of friction and noise in
the factmaking process.[57] Like the supposed neutrality of online platforms,[58] such depictions leverage long-standing beliefs in technology as
a pan-contextual solution;[59] take any process, add technology to make
it bigger, easier, and faster. But Marx had already understood that the
introduction of any such technological acceleration will, in the case of
labor, entail a transformation of social relations between worker and
worker, worker and capitalist.[60] Technology is the material means by
which newly modified relations between human beings—and, in our
case, between the subject and their own "self"—are concretized and
made conventional. The language of neutrality overshadows what is in
reality a redistribution of credibility, responsibility, labor, in the work of

knowing the self—a reshuffling that leaves the thinking, feeling subject somewhat in the cold.

In this sense, self-surveillance is founded on the privileging of a certain *machinic sensibility*: the persistent and automated activity of collection, analysis, and prediction that establishes a rich communicative circuit between the tracking machine and the body that speaks. In other words, the smart machine's claim to superior knowledge is predicated on its ability to plumb the body *sans* the interference of the human subject. Some types of measurements, such as galvanic skin response, are absolutely beyond human access; they are signals emitted by the body that human sensibility and cognition have no equipment to perceive. In other cases, such as the number of steps taken, the sheer frequency, precision, and volume of the recording process place the data beyond human capabilities. The result is formally analogous to the effect of "black-boxed" algorithms on social media platforms and state surveillance systems.[61] The many micro-judgments that go into what counts as a step, what counts as "good" sleep, are increasingly placed outside the subject's reach (or even retroactive evaluation). As self-surveillance becomes increasingly commercialized, the selection of biophysical markers to be measured and interpreted becomes determined by a combination of what machines are able to sense in the world and what delivers the most spectacular and attractive visuals of diagnosis and improvement. Increasingly, individuals "do not have a lock on which contexts and levels are able to generate meaning" about themselves.[62]

Such a mediated relationship extends and transforms epistemic processes found in predigital forms of self-surveillance. Consider a well-known example of a precomputational technique. In 1726, the young Benjamin Franklin devised for himself a schema of thirteen virtues—ranging from temperance in food and drink to industry in efficient labor.[63] In his diaries, he constructed simple tables of the virtues and the days, where each night he would mark his observance (or lack thereof) of a given virtue: a discrete and ritualized process presided over by the reflexive subject. In such a process, the subject's conscious mind, and human memory, grapples with the day's squabble with a neighbor or the culinary temptations in the evening fare, and it is the subject who confronts the rigid and unyielding table with the devil in the details.

Today, similar efforts to track the self across the everyday entail a rather different experience of the data. An illustrative, if singular, example comes from the QSer Tahl Milburn, who designed and installed what he calls a "Life Automation System" (LIAM) in his own home.[64] There, USB sticks and other objects glow ambiently with colors corresponding to a single "LifeScore" derived from personally tailored and weighted variables. The score aggregates Milburn's net worth, the market performance of his investments, weight, activity, sleep, age, and more. Here, the subject's perception *of* sensing machines itself becomes atmospheric rather than discrete; meanwhile, LIAM ceaselessly communicates with Milburn's own body as well as a host of other machines to generate the feedback. The self becomes "known" at a level that the subject cannot actively track. This phenomenological disjuncture becomes precisely the space for not just overriding human sensibility as a source of information but manipulating it to engineer desired psychophysical states.

Whereas systems such as LIAM remain, for now, the preserve of the few with sufficient technological acumen and the economic means for such elaborate installations, a similar shift is already occurring in cheaply and widely distributed applications. Quantrepreneurs quickly tried their hands at not only health and productivity but also everything from friendships to mood swings and even sex. The app Spreadsheets, as we saw in chapter 1, offered to track users' sexual behavior, including measurements of duration, number of thrusts, and loudness. These metrics clearly were not the result of any philosophical reflection on what is worth measuring about sex; rather, they came from what one could measure quickly and easily, given the kinds of sensors available on a typical smartphone. But over time, what is legible begins to take a practical priority over what remains unreadable by machines. The beauty and terror of this kind of epistemology are that it will produce conclusions purely based on what it measures and, in doing so, suggest that all it does not measure cannot overrule the correlation it has discovered.

All this is not to say that we pass simply from a human-centric model of self-knowledge to a machine-centric one. The division between human and machinic sensibility cannot be an absolute one, and neither can the alleged rise of the machinic over the human be so complete. But there is a shift in the balance of what kinds of external sources speak the truth of our bodies. Even as cultural imaginations of frictionless control and augmentation motivate wearable technologies,[65] the growing veridical authority of technical objects reintroduces the problem of the gap: the injunction to know beyond the limits of human phenomenology and the presentation of autonomous machines that may bypass human subjects and thus overcome their limits. Surveillance "in here" thereby reprises the political organization of ignorance "out there."

Data's Privilege

The human responses to this privileging of machinic sensibility tell us much about the transformation of social relations surrounding self-knowledge. Who (or what) speaks for whom in which contexts verifies the truth of the body and takes responsibility for that truth? The advent of machinic sensibility was not universally celebrated. Bullish proclamations for data's intimacy prompted public backlash, reprising well-worn binaries of human and machine, nature and technology. Mainstream media often played host to this familiar back-and-forth:[66]

> As beneficial as a trove of personal data can be, though, there are some things better left uncharted. Comparing our friend counts and vacations with others on Facebook is already making us sad. And it's unlikely that comparing our lovers' average duration and decibel volume to others' is going to make us happy. Analytics are creeping into the most intimate and unquantifiable parts of our lives. . . . "Filling up a love tank isn't the same as having a personal connection."

Such rhetoric presumes a certain kind of common sense, a set of shared intuitions about what is "natural" to humanity. Notably, such arguments were often grounded in what they left unspecified. How is this "personal connection" incompatible with analytics? What makes sex "unquantifiable," exactly? The refutation of self-surveillance's fabrications often resorted to an ideal picture of natural human being-in-the-world. In attacking the promises as fantastic tales of posthuman cyborgs, they themselves would reify a mythical figure of the pretechnical human and their "natural" relationships. The Romantic legacy survives strongly in these sentiments—that technology is outpacing the human ability to comprehend and manage it and the recurring trope of being watched by invisible forces.[67]

In this reactionary discourse, one key difference between human/ nature and machine/technology was to be quantification. Quantification, of course, generates a powerfully efficient grid of intelligibility— but at a cost. The decision of what to count and how to count it requires dividing the phenomena in question into discrete, measurable pieces now amenable to decontextualized circulation.[68] Criticism against self-surveillance's love of numbers took up two general aspects of this problem. First, it was alleged that the focus on numerical measurement privileges that which is countable over that which is not.[69] The majority of self-surveillance solutions in this period relied on finding acceptable values that might serve as a proxy of the actual phenomena of interest, from the seemingly reasonable (heart rate variability as a proxy of stress) to the incredulous (the aforementioned thrusts per minute). To be sure, the mere use of a proxy does not disqualify the measurement, but skeptics sought to show how self-surveillance's promise of data-driven certainty produces its own margins of uncertainty, its own dark regions.

Quantification was also accused of cultivating a pernicious expectation of numbers-driven optimization. Smart machines' numbers provide discrete and manipulable correlations: I walk more; my steps count goes up. Where enthusiasts saw the golden gate to self-improvement, skeptics argued that such an interface produces a misleading sense of control: the tracker does not become "healthier" in the fuller sense but does become very good at optimizing their Fitbit numbers.[70] Critics warned that the proliferation of data was beginning to masquerade for legitimate and reliable "knowledge."[71] Technology critic Evgeny Morozov joined the fray, reviving a 1987 slogan from philosopher Ivan Illich: the "imperialism of numbers." Morozov argued that quantification gives us the easy, "low-hanging fruit" in our quest for self-knowledge and, in doing so, steers us toward a passive solutionism: just let the machines gather, and we will eventually be better off.[72] The attack on quantification organized the debate surrounding self-surveillance into a normative and ethical problem: What kind of self-knowledge should individuals strive towards, and how? If "better" knowledge is not simply a synonym of technological efficiency or precision, how should it be defined?

Throughout, there is a strong sense of vulnerability as a human price of machinic sensibility. Yet I would argue that the truly consequential point of vulnerability is where the tracking individual becomes known to (determined by) other peoples, institutions, and systems of governance behind the machines. It is essential to grasp this systemic element of what was often marketed as a personalizing, and therefore "private," practice. One indication of this disjuncture was the growing consternation of some QS enthusiasts at trends in the wider industry. It was an industry they had helped popularize but now feared was establishing a massified and standardized market that reduces "n = 1" into a vacuous marketing slogan. These connoisseurs, who had often cobbled together do-it-yourself solutions for each and every individual case, could reasonably claim a more complex human–machine relation in which the tracking subject had a certain personal mastery over the means by which they are datafied.

The 2015 QS Conference afforded a keynote slot for two longtime QSers: Dawn Nafus, an Intel Labs anthropologist, and Anne Wright from Carnegie Mellon University's Robotics Institute. They warned the audience of a "tyranny of the norm," where my personal reflection of my data is shaped by the debilitating idea that the healthy and happy normal is a normal value on the populational curve. Gary Wolf, nodding from the sidelines, shared his consternation at a recent tracking product that promised to compare each user to "the global norm"[73]—an invitation that, he later told me, he found "severely perilous."[74] The position of the individual, decentralized, tinkering self-tracker in the face of a massifying industry mirrored, in some senses, the threatened position of the active, knowing subject as the production of self-knowledge began to institutionalize.

Across QS and the wider public, these criticisms reprised classic humanist positions (to varying degrees of sophistication), pitting the integrity and agency of the human subject against not only the machine but also the social relations of systematic capture and control that the machine represented. We find here a remixing of two familiar figures:

1. The hijacked body. The body—and through it, the mind, the subject, "identity"—is overtaken by external forces through technological means. This is the recurrent fantasy, both seductive

and terrifying, in popular science-fiction and the cybernetic imagination.

2. The exposed[75] body. The body's control over its own boundaries of visibility are violated, and what it would like to keep secret, to disavow, to leave underdetermined are forced out—not only against their will but, more often, *oblivious* to their will as well. Again, we find such nightmares to be a staple in science fiction (and other hypothetical descriptions of technology out of control) from at least the mid-twentieth century.

The two figures share an essential continuity: the hijacked body is the subject's loss of mastery over action, the exposed body the loss of mastery over perception and knowledge. I become known against my will, and I am made to act (i.e., produce new truths) against my will. They express the fear that the conventions of differential visibility and control that we typically shorthand as "agency" or "privacy" are collapsing. Recalling Nagel, it is also that the ways in which we made a virtue of concealment are now subject to a unitary insistence on objective truth. Subjects are urged to make themselves as transparent as they wish of their governments.

It was precisely these underdetermined areas—full of relations and affects with which society has previously chosen to practice nonacknowledgment—that became lightning rods for the imagined conflict of human and machine. Sex, as we have seen, was a reliable provocateur. kGoal, a crowdfunded "Fitbit for your vagina," promised to track pelvic muscle training—and was roundly criticized again as "quantifying gone too far."[76] Even the bowels were not immune to data's colonizing principle, with Poo Keeper offering to help users keep track of fecal matter. Human relationships and individuals' psychic lives were also subject to figurations of exposure and hijacking. In 2015 came pplkpr: a smartphone app that promises to automatically track the emotional impact of one's human acquaintances. The GPS data on the smartphone and heart rate variability measurements via a wristband are combined with prompts asking users to identify who they had just met and how the interaction made them feel. The data is used to deliver correlations about which acquaintances might make the user more nervous, aroused, happy. The promotional trailer proclaimed,[77]

See how your relations stack up, and let pplkpr find the ones that work for you. It'll automatically manage your relationships so that you don't have to, scheduling time with people that make you feel good, and blocking the ones that don't. Forget fake friends, failed romance, and FOMO [fear of missing out]. Optimise your social life with pplkpr.

pplkpr presented user testimonies that point to the kind of veridical authority commanded by data's claim to objectivity. "Using the app as a justification for not wanting to spend time with someone is a lot more definitive than just saying, I'm uncomfortable," said one. "It made me realise the truth," said another. In this case, pplkpr's creators had preempted the criticism: as artists-in-residence at Carnegie Mellon University, they claimed that the app is both a genuine tracking solution and an "art project" and that they themselves are "really critical of the current attitude towards [quantification], which seems to be super utopian, data happy, collect everything."[78] Again, this criticism was directed at the idea that what is best left nonacknowledged, or amenable to individual control, was becoming objectified, standardized, and publicized.

These anxieties were again connected to the massification of self-surveillance and the growing networks of social relations governing this newly produced knowledge. Far more earnest—and controversial—than pplkpr was Peeple, an app that was announced in late 2015 promising a fully public system for quantitatively rating ordinary humans. Individuals would be broken down into romantic, professional, and personal "scores" as rated by their acquaintances—and they would not be able to opt out. Technically more proximate to social network platforms than individualized trackers, Peeple demonstrates the overlapping capacities for and concerns over knowledge of the self and the social relations established through them. Brought to public attention through *The Washington Post*, Peeple set off a firestorm of international controversy.[79] The impending app was quickly branded a "creepy" poster child for a "rating" society.[80] Peeple would ultimately be released in an emaciated form, lacking many of the most intrusive—and therefore useful—features. Yet soon after, China would begin experimenting with Zhima Credit, which combined elements of traditional credit scoring with the Peeple-like aim of producing public scores on individuals' social, moral, and human qualities. The boundaries between localized self-surveillance and the wider data market's demand for ubiquitous measurement would prove difficult to maintain; after all, adhering to such boundaries would be to contradict the data market's fundamental impetus for profit and control.

One area in which this cross-contextual recombination of tracking data is making headway is the juridico-legal. In 2015, one Richard Dabate reported a murder; a masked intruder had entered the home and killed his wife. The police arrested Dabate instead. The warrant application for Dabate shows that the detectives found Dabate's testimony suspicious and subsequently turned to various archives of Dabate and the victim's communications: texts, phone calls, email, Facebook activity, Apple iMessage, and, notably, the victim's Fitbit activity tracker. Its record of her physical activity, released to the police by the wearables company, suggested that the victim was still alive and moving well after the suspect claimed he had seen her shot to death. These measurements, although not conclusive on their own, contributed to the police's rejection of Dabate's narrative and his arrest on suspicion of murder.[81] The Dabate case was not unique, either. Months later, an Australian murder case took a remarkably similar turn when Apple Watch records contradicted Caroline Nilsson's claim that her mother-in-law had been killed by intruders. In 2014, the popular fitness tracker Fitbit had already been called as a legal witness. A Canadian law firm invoked the client's own tracking data to prove that a work injury had affected her physical activity levels adversely.[82] Such cases point to a near future where a datafied "me" might appear in court and establish an authority over the truth of the self that can override my own words and memories.

The Quantified Us

As these devices fabricated new information about human bodies, these facts began to come back and shape how bodies count in the social world. It began as the Quantified Self; increasingly, the industry would speak of the Quantified Us.[83]

Consider Fitbit, the poster child of self-surveillance's mainstream popularity. Having found early success by enticing individual consumers to track their fitness, the company proceeded to court corporate clients as well. By the mid-2010s, companies such as John Hancock (US) and Sovereign (New Zealand) had begun experimenting with subsidizing

Fitbits for insurance customers and offering gamified rewards for sharing their exercise and movement data.[84] The insurance industry also looked elsewhere for new sources of data on the health and risk factors of its clients; American companies such as Progressive and State Farm were found offering drivers discounted rates for exposing data logged in their cars' telemetric units.[85] By 2018, John Hancock had pivoted to selling exclusively "interactive" policies, integrating health tracking data into all of its products.[86] For the moment, such partnerships made clear that individuals' fitness data would not be used to recalculate their insurance premiums. Yet exactly the horizon of potential use motivates such business partnerships in the first place.

These efforts are part of a broader mobilization of new tracking technologies for existing strategies to manage labor power.[87] Fitbit and other exercise-tracking devices are also being used by employers such as BP America, extrapolating productivity and other value judgments on the basis of movement data.[88] In such arrangements, workers' bodies are embedded with an array of tracking machines that report to another master. In 2018, Amazon registered new patents for customized wristbands for its warehouse workers; again, the immediate purpose remained modest, confined to helping the workers locate the right inventory bin as they handle items for shipping (figure 4.2). This mundane suggestion, however, must be placed in the context of a workplace that is one of the most profitable centers for mechanizing human labor in the internet age. Amazon's warehouse employees have been reported to suffer from conditions including extreme heat, abrupt terminations, long hours, and insecure "zero-hour" contracts. Here, the efficient extraction of human labor coincides with the commercial adaptation of tracking technologies; at least since 2013, Amazon warehouses in Britain had already featured tracking equipment for its workers, monitoring each individual for the rate of package processing.

If self-surveillance was initially sold to the public on the basis of future projections about its empowering potential, it is equally clear that the technology was also animated by a vision of exhaustive transparency—this time, of individuals to corporations and economic calculations—as a path to monetization. Part of the datafication's attractiveness, as I have mentioned, is the idea of frictionless scaling. Just as Facebook went from Harvard dorms to the world, so, too, would each tracking solu-

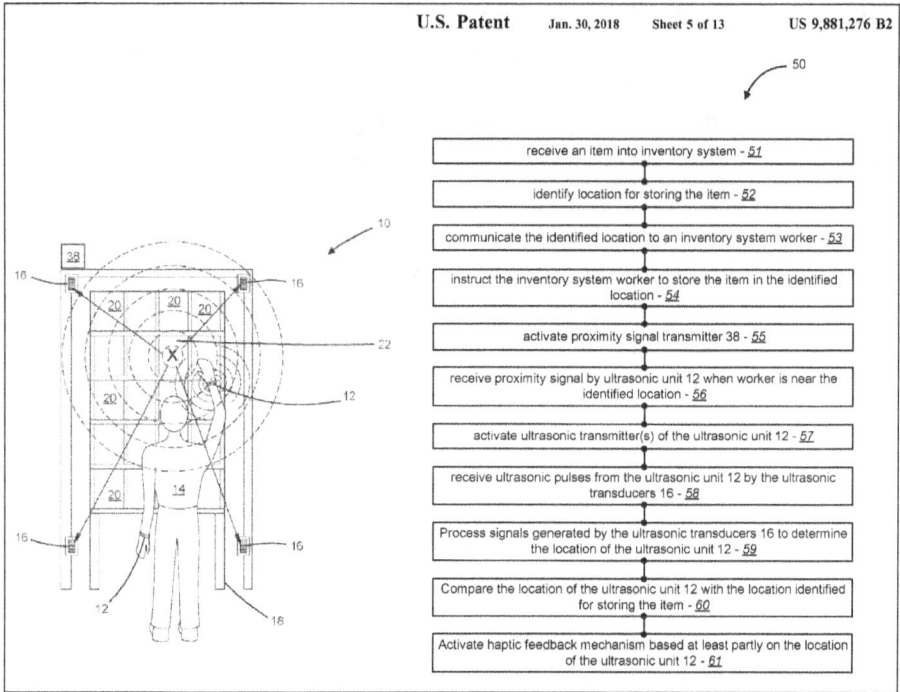

Figures 4.2a and b. Amazon's patent describes an ultrasonic wristband for warehouse workers, which would detect proximity to the right inventory bin and alert the wearer. Source: Jonathan Evan Cohn, Ultrasonic Bracelet and Receiver for Detecting Position in 2D Plane. US 9881276 B2, USA, issued 2018.

tion dream of global market penetration. During this upscaling process, the ostensible democratization of datafication technologies swings back round to furnish new possibilities for recombinant control. For instance, standardized tracking of a group of workers means that data about fellow workers may be brought to bear on myself, even if I might have avoided direct capture. Going further, groups are able to be managed through data collected about other groups; one study shows that retailers have used customer data to "refractively surveil" their workers—circumventing any legal restrictions that might be in place for directly surveilling employees.[89]

Two of the most active areas for this "control creep" were health and wellness—areas that were also quick to invest heavily in this generation of self-surveillance technologies, often carrying labels such as mHealth and e-Health. Some early offerings sought to provide existing health care systems with more comprehensive surveillance over their patients; skin patches or bandages might be equipped to communicate glucose levels or heart rate fluctuations to doctors in real time (or close to it). By the mid-2010s, the largest tech corporations, including Apple, were pouring funds and manpower into exploring more comprehensive data sharing across health care institutions, individual users, and, of course, the corporations themselves. In 2017, a San Francisco–based start-up named Clover Health won the coveted unicorn status—the nickname bestowed to those achieving more than 1 billion USD in valuation—partly with venture capital funding from Google Ventures. Born in 2014 as a company offering medical insurance, Clover ballooned to unicorn size by promising to advance the turn to "patient-centered analytics" through predictive systems. Although still dressed in the more traditional language of medicine, doctors, and insurance programs, Clover's pitch is that its software will draw on anomalies (such as failing to pick up a prescription) to predict issues and then follow up by calling the client or arranging a home visit. This relies on Clover's ability to collect and triangulate a large variety of data, including demographic and payment information, tax identification, and medical records of patients, including consultations and treatment plans.[90] As firms such as Palantir enable the circulation of data and technology between the state and the private sector, companies such as Clover are brokering new connections between self-tracking and workplace management, between institutionalized health care and the data analytics of "n = 1."

Such upscaling is not frictionless. For one thing, the actual circulation of personal data is typically constrained by laws and norms. Here, traditional safeguards such as individual consent hold sway—another artifact of the liberal subjectivity that is supposed to tightly control everything that has to do with their "selves." But such quaint assurances of privacy are becoming increasingly ineffective. In 2017, the US Food and Drug Administration approved Abilify, a pill fitted with sensors for tracking intake. It claimed that patients would be able to check their ingestion on their smartphone and "permit" caregivers and physicians to access the information. The medicine in question was aripiprazole, typically administered to schizophrenic, bipolar, and depressive conditions. Yet what would it mean for an underage or a senior patient diagnosed with a

mental disorder to "refuse" consent for doctors to access medical information? What is the weight of choice available once such forms of data
sharing become normalized? And what of cases where consent is violated at a mass scale? One well-known example involves MEDbase 200,
a data broker that took lists such as rape victims and erectile dysfunction
sufferers from aggregated medical records and sold them for years before authorities noticed.[91] The commercial incentive to "scale" technologies of datafication will continue to dilute the practical value of privacy
as a discrete choice. As it has been noted with social media platforms,
"the price of nonparticipation" is often amorphous, while its benefits
are clearly articulated in instrumental language.[92] As long as consent
remains an act of choice, it remains theoretically possible to withdraw
from this intense circulation of data. Yet what kind of individuals can
afford to pay the price, both tangible and intangible?

What began under the auspices of individuals that know themselves
gradually expands into new forms of institutional power over the conditions of our social existence—traversing social and cultural barriers via
the logic of surveillance capitalism. During the 2000s and 2010s, the
very promise of empowerment through online connectedness coaxed
billions of users to voluntarily opt into the relentless surveillance and
exploitation of their personal data. The convenience of "connectedness"
had as its price a "connectivity" to vast corporate surveillance systems.[93]
Users only belatedly realized that "if you're not paying for it . . . you're
the product being sold"[94]—not only through the direct monetization of
personal data for advertising but also in users' enrollment into general
systems of social sorting for more *efficient* determination of individual
truths.[95] In the same way, self-surveillance technology is beginning to
coalesce into a wider ecosystem for the production of my truth by others, and in ways that I cannot follow.

The Fitter Self

This ongoing integration of individualized self-surveillance with the
wider data market, and the institutionalized production of bodies into
facts, interpellates a particular kind of human subject. If state surveillance focused on the figure of the lone wolf, the dangerously unknown
subject, then here we find a virtuous (i.e., more amenable to capitalism

and computation) counterpart: the tech-savvy, self-caring, well opti-mized, fitter self.

In 2014, several media outlets ran stories on what they called the "most connected man on earth."[96] Chris Dancy wore augmented vision glasses and monitored home air quality. His tracking devices included heart monitors, posture sensors, and a separate system for his dog; a snapshot of his Google Calendar shows a torrent of labels (figure 4.3). The sheer number of devices and data that he would juggle on a daily basis illustrated what it takes to build one's everyday life around these measurements: the money, the time, the tech literacy, the living envi-ronment. Although self-surveillance is typically described as lifestyle consumption, it is also a form of labor: the work of optimizing every aspect of one's life, to achieve maximum productivity, fitness, and hap-piness. It is a kind of labor that has no end point, a labor that occurs 24/7 and constantly mixes the boundaries of public and private, work and leisure. Health and productivity might be considered natural things for any individual to pursue. Yet the late capitalist worker is increasingly compelled to invest in relentless self-optimization to mitigate their con-ditions of economic precarity. A willingness to measure oneself against metrics of wellness or productivity often dovetails with highly moralized social expectations about what it takes to be a successful worker—or just one that can land a job.[97] And if such self-surveillance is costly, then their increasing normalization means that the costs of failing to datafy oneself are also rising. A few years later, Dancy himself would muse,[98]

> It costs a lot not to have internet access. I wonder in the future whether people will be able to afford not living with the internet . . . Not being connected will constitute a new disability.

This fitter self is not born *ex nihilo* but extends a longer history of expec-tations and idealizations. Faced with market pressures that demand a more optimized, more machinelike human subject, the worker has long been asked to take up the indefinite and unpaid labor of making one-self competitive. In this regard, self-surveillance extends the work done by wellness and health in preceding decades—domains that governed how subjects were to understand and care for their bodies. During much of the twentieth century, Freud's popularity provided a conduit for the

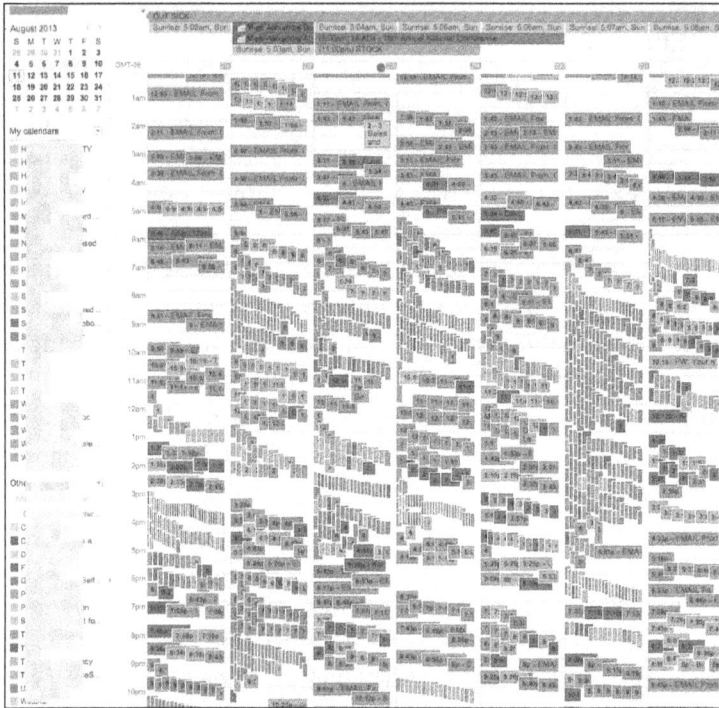

Figure 4.3. A Google calendar feed of some of Dancy's data.
Source: Sarah Griffiths, "Is This the Most Connected Human on the Planet? Man Is Wired up to 700 Sensors to Capture Every Single Detail of His Existence," *Daily Mail*, March 25, 2014, http://www.dailymail.co.uk/sciencetech/article-2588779/Is-connected-man-planet-Man-wired-700-devices-capture-single-existence.html.

application of therapeutic and psychoanalytic knowledge regimes over areas such as self-help, workplace management, and marriage advice.[99] As with self-surveillance, these were areas where the introduction of external systems for knowledge production brought together relatively "private" domains of human life with external and public systems of knowledge production.[100] Self-surveillance's visions of transcending biological limits of self-knowledge, and of aligning machinic sensibility with human cognition, complemented older projects, such as the "Silk Road" traffic of imported Eastern meditative practices since the 1960s.[101] These techniques were also united in their designated nemesis: the psychophysical malaises inherent in the modern subject, carrying various labels from stress to fatigue to information overload. Health as a question of optimization, the subject as *responsible* for the body, the emphasis of the body as a biological and scientized object—each principle is inherited from these existing rationalities of health and wellness.[102]

Wellness, in particular, had been construed as a problem of willpower and self-discipline ever since its emergence as a subcultural neologism.[103] In 1978, *New York Magazine* ran a cover story on "The Physical Elite," sketching the essential stylings of the wellness subject: fastidious self-control glossed with mysticism, conspicuously consumed lifestyle products dunked in New Age spiritualism. Demographically, they were "the 'upscale' people, college-educated . . . upper and upper-middle income brackets, professionals, business people, the general run of white-collar workers."[104] Forty years on, the core audience of self-surveillance technology appears remarkably similar—although infused with a high concentration of tech industry professionals and, reflecting the industry writ large, a heavy male bias.[105] Especially in New York and Silicon Valley, two of the liveliest epicenters of QS activity, many turned to self-surveillance alongside yoga and meditation. The spiritual and reflective dimension would hopefully fill up what one tracker termed the "question mark in the middle" between objective measurements and its promises of human transformation.[106] Where the counterculture (then the mainstream) took up New Age spirituality as an exotic route for consuming oneself out of capitalism,[107] datafication's synergy with mindfulness again offers technological consumption as the antidote to technology's discontents.

The ideal subject of self-surveillance further inherits from wellness a certain rationale of willpower and discipline. Gary Wolf argues that "we are in no position to stand guard over our judgments without the help of machines to keep us steady."[108] Such rhetoric resonates with historical examples like alcoholism, long characterized as a "disease of the will."[109] Twentieth-century conceptions of alcoholism, and programs to combat it (chief among them Alcoholics Anonymous), established the necessity of external discipline to help steady the will and thereby deliver a wholesome subject. The pattern would be replicated in later generations' focus on dieting, exercise, drug addiction, and, indeed, self-surveillance. The introduction of autonomous tracking technologies again prompted declarations of the chronic weakness of will in humans, and the necessity of machinic supplements to achieve a healthy level.[110] Of course, we should not mistake machinic sensibility as operating freely on a passive subject. Rather, the alleged deficiencies of human willpower are leveraged to compel voluntary acceptance of techniques for externalized determination of the "good" body. Foucault tells us that essential to discipline is a moralizing element.[111] It encourages subjects to idealize the conditions of their own disciplining and enshrine the goals it proposes as virtues that are ends in themselves. In the case of wellness, the very concept helped extend the ideals of health beyond the lack of malfunctioning body parts to an optimal state of fitness and energy, packaging the biopolitical ideal of productive manpower in terms of individual happiness and self-realization.

Labor, Discipline, Morality

Accordingly, self-surveillance technologies today are conceptualized as tools not only for observation and prediction but also for recommendation, curation, nudging, and habituation—a focus especially pronounced, once again, in practices of tracking for health and wellness.[112] This trajectory is well summarized by Vinod Khosla, a tech industry veteran who had co-founded Sun Microsystems in the 1980s and has emerged as a key figure and investor in the self-surveillance industry. "Humans look for what they know to look for. The next generation of algorithms will look for everything"—and then tell us what we did not know to look for, to boot.[113] Other actors in the tracking business similarly spoke of using tracking to achieve a "planned serendipity," where the machinic system furnishes the moment you are to seize.[114] This vision was one with broad traction across the data market: Google's evolution from a search engine giant to a primarily artificial intelligence company hinges around a similar ambition, memorably captured in then CEO Eric Schmidt's viral comment that "people don't want Google to answer their questions, they want Google to tell them what they should be doing next."[115] As tracking technologies monitor and manipulate human living at a bodily and preconscious level, they shape new "somatic niches": modified rhythms of sleep, comfort, and attention, all the ways in which human subjects are always already situated via their bodies.[116] Just as the path to a sober self required submission to externally curated exercises of willpower, self-surveillance promises a subject who truly knows oneself free from comforting lies or faulty memory—once they willingly sign up for systems of persistent monitoring and calculation.

Caught between the data hunger of the market and the moralization of self-knowledge, an increasingly complex and persistent form of labor becomes demanded of the tracking subject. Historically, ideas of especially psychological wellness in the twentieth century put ever greater pressure on ordinary subjects to know and manage their emotions—a task that, in turn, legitimized the expense of money and trust on emerging therapeutic techniques.[117] Even as new technological solutions promise to relieve us of this overbearing labor, they contribute to the normalization of the expectation that the healthy, responsible, productive subject knows and cares for oneself in these complicated ways. Chapter 2 showed how the valorization of transparency burdens the public with an impossible labor to "know for themselves." This gap in knowability, furthermore, does not erode the promises of better knowledge through technology but cements big data's place as necessary for the challenges of contemporary societies. Self-surveillance's gradual evolution from the optional experiments of connoisseurs to a broader norm reprises this ambivalence. As Melissa Gregg has shown, work in many contexts has steadily grown more "intimate," requiring emotional labor, dominating our social and personal identity, and "bleeding" into every space and time of ordinary life.[118] The emerging traffic of techniques and information between the tracking we do unto ourselves and the tracking that defines us as workers is the next frontier of such precarity.[119]

Schematically, we might identify four related aspects that constitute the techniques of subjection embedded in self-surveillance practices:

1. *An aspirational trajectory.* To be subject is to be made to desire, engaged in a becoming towards *x*; interpellation does not call out the individual as what *is* but initiates a becoming.[120] The irony of data's intimacy is that the more we measure the present, the more we end up chewing over the past and racing after the future. At the end of the tunnel stands a certain ideal of knowledge and care: a (mythical) individual whose knowledge is fully aligned to empirical, scientific truth and thereby achieves felt harmony with one's own body. In the messy world of practice, the anticipatory,

predictive sensibilities of self-surveillance produce not a singular identity but a multiplying set of interfaces and points of comparison. The fabrications do not always cohere into a clear body but remain a disparate array of measurements and "insights." I become known *in relation to* others (the "global norm"), to the curvature of the numbers, their extrapolation into the future. The individual is corralled into an indefinite relation of consumption, guidance, examination, recording, and judgment.

2. *Consumerist discipline.* Aspects of Foucauldian discipline are remixed into the apparatus of "lifestyle" consumption. Its objective—the cultivation of optimal productivity—intersects a larger biopolitical function with the self-interest of the tracking individual.[121] Especially for urban, tech-savvy populations of early twenty-first-century America, this active and competitive management of productivity across the entire bandwidth of everyday living emerges as a profitable strategy in the face of increasingly "entrepreneurial"—mobile, high turnover, flexible, project-based— labor conditions.

   Marx understood that the capitalist does not simply purchase the worker's labor power but also invisibilizes the nature of this relation to derive a profitable margin.[122] Technologies of self-surveillance abstract away the many costs of datafying the self, presenting a bargain in which the user has everything to gain for the simple price of the gadget. Meanwhile, the carrot of empowerment and self-realization is accompanied by the stick of the market: Can you really afford to stay unoptimized in the age of irregular employment and artificial intelligence? Conversely, if you should accept the invitation, can you afford the hardware, the technical knowledge, the time to experiment, and the kind of lifestyle and living conditions compatible with intensive tracking for self-optimization?

3. *The moralization of predictivity.* Although these aspirations and disciplinary techniques generally receive a moralizing gloss, one particular consequence is the valorization of predictivity. This entails not only the laudation of quantified data as privileged paths to self-knowledge but also a renewed embrace of consistency and regularity as human virtues. Here, self-surveillance hearkens back

to clock time and other modern technologies for organizing and quantifying human behavior. Such technologies helped establish measurable, discrete temporality as a benchmark for not only how productive workers were but also their moral qualities. A punctual worker was a worker fully in sync with the machines that increasingly inhabited their workplace, a worker whom the capitalist could trust to function with mechanical regularity (the same average output on the factory line, hour after hour).[123] In both Google's bid to "tell [users] what they should be doing next"[124] and self-surveillance's promise of anticipating what habits and practices are "best for you," we find the attempted organization of every human quirk and tendency into the kind of data that can be subsequently recombined for purposes as diverse as insurance, employee monitoring, and criminal investigations.

4. *The problem of the body.* The Sisyphean irony is that the more we aspire toward the harmonious mastery that is the prize of objective self-knowledge, the more we become entangled in a persistent, indefinite identification of the body as a problem. The body becomes an entity to be constantly questioned, analyzed, regularized, intervened, and optimized. Wellness, one early twenty-first-century guru claims, is about "solution-oriented people":[125] folk who are not content with mediocre living but are intent on feeling the best they possibly can. Indeed, it is the quest for solutions to specific personal problems, like chronic health conditions, which often brings individuals to self-surveillance. Unlike a simple conception of "health" as the *lack* of illness and disease, in which the ideal body is the invisible body that lets the mind do what it wills, the body of wellness and the body of self-surveillance are ones that must always continue to be worked on without a clear endpoint, extending a long historical shift partly rooted in the pharmaceutical industry's own efforts at expanding its market.[126]

Each of these elements emphasizes the ever-deepening intersections between the phenomenological and the structural. The privileges granted to machinic sensibility transform what the individual is able to know about themselves; what they must know about themselves to be normal, optimal; the information they must disclose to become intelligible to society. The rhetoric of *self*-tracking, *self*-knowledge, obscures the critical shift in the balance of what the human subject can say about themselves, and what is said about them by others.

In 2014, Alex Pentland's *Social Physics* hit the best-selling lists. Pentland, director at MIT's Media Lab and dubbed the "presiding genius of the Big Data revolution" in the book's promo blurbs, showcased an array of new technologies for tracking human behavior and sociability to, as the book's website put it, "create organizations and governments that are cooperative, productive, and creative." One key item was the sociometric badge. Worn by employees, it would collect data such as voice tone, body language, and conversation patterns (who talks to whom). Katherine Hayles has called this kind of data gathering "somatic surveillance": the exteriorization of what previously remained nonrational and nonconscious, making it available for conscious (and collective) consideration.[127] Such "interior depths" had always been sought after, of course. But recall the techniques of François Leuret. Whereas torture or confession relied on the conscious subject to mediate and validate whatever was being extracted, this is no longer the case at all in the world of sociometry. The individual need say nothing; even in court, the Fitbit has already spoken on their behalf—or against them.

The sociometric badge is no waterboarding. But across the sociometric badge, Fitbit's alliance with insurance companies, "human out of loop" systems for calculating credit scores,[128] or "threat scores" used to predict and profile potential criminals,[129] the question becomes: What new relations of vulnerability are produced through these additional ways in which we become spoken for by others? What new labors for making ourselves consistent, respectable, socially normal, profitable, and legible become levied on the self turned inside out? What is at stake is not the continuum of influence between humans and machines, the problem of machines taking over what humans had in a pretechnological status quo. The key difference lies in the position of the conscious, experiencing subject relative to their own body and social identity: a relation that had never been completely internal and had always been fraught with vulnerability. As one critique of self-surveillance put it, "'becoming oneself' has turned into a crappy job—a compulsory low-paying, low-skill job."[130] A job whose labor entails not the experience of discovering the inner authentic self but a production work whose results may no longer belong to myself, either.

As technologies of self-surveillance scale up to a wider data market, these questions increasingly apply far beyond those that willingly (or, at least, knowingly) subscribe to tracking. Just as state surveillance constantly divides society into different kinds of bodies, self-surveillance also produces implicit divisions between the fitter selves and the rest—thus reintroducing enduring social and economic inequalities. In 2017, Y Combinator, one of tech's most prestigious accelerator programs, agreed to fund Flock, a safety start-up. Flock's product would be cameras installed on private property, which would then film vehicles passing through the neighborhood. The data could then be shared with the police in the event of a crime. The technology offers a private supplement to the ongoing expansion of state surveillance systems. Flock was far from alone in pursuing this niche. In 2018, Amazon acquired another company, Ring Inc., and began to aggressively push the home surveillance system through its storefront. Ring explicitly theorizes itself as a product linking state and self-surveillance: Jamie Siminoff, its CEO, emailed employees in 2016 to announce that the company was "going to war with anyone that wants to harm a neighbourhood."[131]

Here, the technological contraption materializes a certain imagination of the social. On one side of the camera is the homeowner, the tenant, those who belong, and those who own. On the other side is a world of dangerous others requiring constant surveillance without consent—in other words, what we used to call "public space." Meanwhile, the consumer's voluntary participation in this kind of domestic tracking opens them up to new networks of surveillance and data extraction; under Amazon's ownership, Ring has begun to actively share user data with local law enforcement, introducing another path to a public-private data market. Historically, the street has always been a fraught boundary between the public and the private; with tools such as Flock and Ring, it is also a place where this turn to data-driven knowledge shifts the balance of visibility, how different bodies are separated into different sides of this factmaking process. This redistribution of visibility and knowability, of rights and freedoms, is where we turn next.

5

Bodies into Facts

"We knew already." Or at least, that was the refrain in the immediate aftermath of the Snowden leaks. Did he tell us anything we didn't know, asked journalists.[1] "They didn't feel much like revelations," said a director.[2] Did this young man risk his life to reveal an unprecedented cache of secret information only for the world to shrug? Snowden himself thought this might be the case. Even as he crossed the point of no return, sending top-secret material to trusted journalists, he wrote of the fear that the public will simply reply, "[W]e assumed this was happening and don't care."[3] But what is meant by this curious phrase, "we knew already"?

"Knew"—yes, some of the information really was public knowledge. *The New York Times* had blown the whistle on warrantless eavesdropping as early as 2005; *USA Today* had followed up with the story of MAINWAY, a National Security Agency (NSA) database of domestic phone call records, in 2006. As we saw in chapter 3, John Poindexter simply showed up to a DARPATech conference in 2002 and revealed the "Total Information Awareness" project to the American public. Snowden, the story went, confirmed what was *virtually known*. But who is this "we"? The discourse designated a depersonalized hivemind: knowledge of NSA surveillance was stored in our collective archive, although the proof was in nonhuman documents rather than what individuals can personally remember. At other times, however, the "we" does become more specific. It designates the journalist, the expert, the activist: the we in the know who pens these commentaries, the we who is less gullible than the average Joe, the we of the we-told-you-so. It is a "we" that is certainly very different from the public writ large, even as such texts invite that public to agree that we knew already. Satire, as it so often does, brings these ambiguities into the open: "We already knew the NSA spies on us. We already know everything. Everything is boring," opined a spoof piece.[4]

In self-surveillance, the experiential question of "How can I know (for myself)?" is essentially tied up with the question of what becomes socially definable as knowledge. In state surveillance, this troubled figure of the public marks a phantom position: the *we* who supposedly already knows, working speculatively and presumptively to maintain a sense of a rational society. At stake here is the distribution of the responsibilities and rights associated with knowing—across not only humans and machines but also humans and institutions, persons interpellated as individuals and as publics, citizens and the state. Who is the "we" who knows already? Who is the subject asked to vocalize the "we" without "knowing already," and who signs off on this knowledge?

This chapter examines how fabrications achieve the status of knowledge, under what genres of justification, and what kinds of action they empower human actors to undertake. Datafication's public presentation may involve a fantasy of purification, but its practical implementations involve inconsistent and locally specific strategies governing how uncertainties are selected and legitimated as predictive insights. To specify this terrain, we turn to three widespread techniques of fabrication that bridge the gap between the alienated public and the ideal public that knows, between the ineliminable uncertainty of terrorist threat and the promised certainty of data. First, subjunctivity describes an "as if" form of reasoning, whereby knowledge and proof are deferred to a future or otherwise hypothetical state. Intimately related is interpassivity: the idea that one does not know, or has not done anything wrong, but "someone else"—or even some*thing* else—has in their stead. Finally, zero-degree risk returns to the fixation with numbers and statistics that we observed in chapter 2. It suggests that the *performative* quality of risk discourse and statistical reasoning contributes to the normalization of "acceptable levels" of uncertainty and speculation. Here, the focus is less on specific algorithms and their internal calculations, or the secretive hyperobjectivity of the surveillance system, and more on the network of expectations, justifications, and imaginations surrounding state surveillance.

These techniques describe, respectively, three sources of epistemic authority that seek to mitigate the proliferation of uncertainty: potentiality, the Other, and numbers. To be sure, modern regimes of knowledge have always entertained a host of heuristics to stabilize its truth claims, to provide grounding. The point is that while flying the banner of machinic certainty, technologies of datafication often create their own dependencies on ifs and maybes. From the FBI's turn toward actively manufacturing terrorist suspects to surveillance discourse's shedding the pretense of calculable risk and probabilistic reasoning, the data-driven society engages in a visible turn toward simulated knowledge by state surveillance actors. These specific modulations of information flow and thresholds of truth spell out the political and moral implications of machine-driven factmaking. Techniques such as subjunctivity are best understood not as irrational and deceptive manipulations of knowledge processes but as part of the constant redrawing of normative rules that govern what relations of deferral, simulation, and speculation come to "count" as knowledge.

Subjunctivity

Your rights matter because you never know when you're going to need them. People should be able to pick up the phone
and call their family, should be able to send a text message to
their loved one, buy a book online, without worrying how
this could look to a government possibly years in the future.
—Edward Snowden

I buy fire insurance ever since I retired, the wife and I bought
a house out here and we buy fire insurance every year. Never
had a fire. But I am not gonna quit buying my fire insurance,
same kind of thing.
—James Clapper, US Director of National Intelligence

The distilled formula: you never know, so you must act (or believe, at least).[5] Both the whistleblower and the public face of NSA surveillance speak the language of loomings: threats that are nothing *yet* but are already very much real in their existence as potential.[6] In fact, the two articulations selectively cleave what kinds of uncertainty should count for how we feel, how we reason, about surveillance and terrorism. For Snowden, what is at stake is the social construction of the diffuse and potential harms of exposed data as something that can be felt, something creepy enough to take up space in our deliberations. For Clapper, it is the effort to tether new forms of surveillance back to the venerable tradition of actuarial reason, drawing a sense of naturalness and legitimacy from our habituated disposition toward fires, floods, and robberies. Just as insurance companies in the past successfully normalized highly specific pricing mechanisms for certain types of dangers, Clapper seeks to position terrorism as an objective form of danger whose mitigation is not a question of politics, only reality.

This is the *as-if*, the subjunctive. As a grammatical rule, the subjunctive mood invokes a statement that it explicitly acknowledges as hypothetical or unproven: "If I were . . ." It acknowledges a given thing to be unproved or presently not actualized *and* nevertheless gives it an operational reality for decisions, sentiments, and predictions. Does James Clapper believe there is going to be a fire, or not? That is beside the point: even if the fire may never happen, it is sufficiently disastrous that we must consider it a real basis for judgment and action. Uncertainty is not counted as a penalty on the quality of the proof, a negative lack of information, but as an aspect of social reality in and of itself that demands a (rather certain) response.

Subjunctivity thus involves a kind of positional plasticity, transposing subject positions across different space/time and alternative possibilities. The proliferation of predictions and correlations require subjects to think, feel, and situate themselves in a wider range of possible worlds. The future becomes foregrounded as the site in which technology will finally deliver the truth and in which the truth about technology can be confirmed. Big data analytics, with their particular interest in prediction and Bayesian statistics, have attracted descriptions of *futur antérieur*: the loaning of what "will have been" to "stapl[e] the future to the past."[7] Derrida puts it still more generally: the future (*l'avenir*), as what is to come (*là-venir*), provides a perennial opening out from the confines of the present, an indefinite horizon from which alternative truths may be pilfered for use.[8]

The fungibility of futures, as sites where fantasy may modify the present, is well understood in the domain of science fiction. Samuel R. Delany defines subjunctivity more broadly as "the tension on the thread of meaning that runs between" words, a "blanket indicative tension (or mood) [that] informs the whole series," infusing the situation with a hypothetical presence. These scenarios are furthermore infused with a sense of possibility—insofar as science fiction concerns itself with "events that could have happened," drawing a basic (if often complicated) boundary with fantasy as that which "could not have happened" in our universe.[9] The connection here is not merely thematic; despite science fiction's traditionally lowly position in the cultural hierarchy, there are numerous historical linkages between its speculative theorizing of our future and the social history of computing and the internet. Spacewar, MIT's 1962 video game that emblematized the move toward personal, real-time, "free" cultures of computing,[10] was envisioned at the time as a nonorthodox attempt at applied science fiction.[11] The hacker-troll-activist collective Anonymous hit on the "Guy Fawkes" mask as its symbol through the Hollywood blockbuster *V for Vendetta*,[12] as well as the "Laughing Man" from the Japanese sci-fi manga *Ghost in the Shell* (which, in turn, derived the figure from J. D. Salinger's *The Laughing Man*). We shall see throughout the chapter how narrativizations of totalitarian surveillance, of ubiquitous terror threats, of suspects that just *had to* be stopped before conclusive evidence could be gathered, all draw from different kinds of "fiction" to realize a mysterious and dangerous reality.

In short, subjunctivity often produces gray areas for bestowing speculative and hypothetical reasoning with a disavowed form of veridical authority. Timothy Melley has shown that the many secret institutions, practices, and policies that make up the dark side of the US government—surveillance included—are often leaked into the public sphere through fictional literature and the use of speculation in official public discourse.[13] He argues that the result is a public sphere of "structural irrationality": as truth becomes regularly accessed through fictional detours (that are again explicitly known to be fictional but serious), there is contamination of the public faith in rational proof. The idea is that cinematic representations of political intrigue and shadow governments thus serve not only to "inform" the public in the short term but also risk undermining the wider public faith in institutions, producing cultural imaginaries of deep states and vast conspiracies. Here, I do not want to focus too much on the language of irrationality, if only to avoid a different kind of fiction that idealizes a "normal," properly rational knowledge. The point is that subjunctivity habituates subjects into the manifold connections between speculations about the explicitly uncertain and nonactual, on one hand, and the operationalization of knowledge—that is, turning knowns and certainties into judgment and action—on the other.

These characteristics mark deployments of subjunctivity as highly ritualistic. Rituals have been called "time out of time," or liminal zones.[14] They are moments when society says, "Wait: let us step out of our rules and rhythms of life for a moment," so that they may be renewed and re-affirmed or undergo specific modifications (such as the change in status of an individual member in a rite of passage). In the Snowden affair and the war on terror, we find subjunctive constructions that allow the audience to "play out" totalitarian futures and next terrorist attacks. Here, the boundary between speculation and reality is even more porous. The simple refrain that "you never know" invites the public to suture scenarios and simulations back into their assessment of "reality," thereby reconciling enduring uncertainty about what it is supposed to have "known already."

As If, They Watch

The parallel arguments by Snowden and Clapper demonstrate both the widespread use of subjunctivity to make sense of state surveillance and the different ways in which those uncertain futures can be selected for concretization. Critics of NSA surveillance were keenly aware that the nature of the technology does not lend itself to discrete "smoking gun" incidents (where, say, an irreproachably innocent individual might be specifically spied on and subsequently suffer from tangible harms). They expected the bulk of the public to believe—and thus prepared to counter the argument—that they are safe as long as they have not done anything "wrong." A *New York Times* op-doc "Why Care About the N.S.A.?" opened with a stern warning from political commentator David Sirota:[15]

> NARRATOR: I want to get your response to a few things people typically say who aren't concerned about recent surveillance revelations.
> DAVID SIROTA: Nobody is looking at my stuff anyway, so I don't care? My argument for that is if you don't speak up for everybody's rights,

you better be ready for your own rights to be trampled when you
least expect it. First and foremost, there are so many laws on the
books, there are so many statutes out there, that you actually probably are doing something wrong . . . So when you start saying I'm not
doing anything wrong . . . you better be really sure of that.

By shifting the subject's gaze onto the bureaucratic and technological
depths that almost entirely lie beyond everyday experience, the subject
is divested of the ability to confirm or deny their own safety. Here, the
reality of surveillance is constructed not (only) by recovering concrete
surveillance practices from their recession, but by expanding their virtual dimension into an enormous, totalitarian as-if.

These uncertainties had much to do with the technical and regulatory conditions behind the surveillance systems. Chapter 3 showed
that the NSA's upstream collection systems target not specific human
individuals but data traces defined as "selectors." Any selector, of
course, potentially involves multiple individuals; any email received
by a selector (in this case, an email address) is by default a two-way
communication at least.[16] (There have been instances where individuals would share access to an email account and use unsent email drafts
to communicate in a bid to foil surveillance efforts.) Meanwhile, the
data involved in these communications itself traverses national borders
in complex ways. Although many of the NSA's surveillance activities
are only authorized against foreign targets, this distinction can only be
sustained through a great deal of cleaning and intelligent guesswork.
Americans who travel, who speak to other "foreign entities," or even
who receive an email mentioning a relevant term could be subject to
so-called incidental collection.[17] For the lay public, the upshot of all this
complexity was that nobody could be quite sure which law protected
them from which kinds of surveillance. Americans' phone records, for
instance, seemed vulnerable to such incidental collection.[18] Tools originally developed for counterterrorism efforts would also find their way
to domestic law enforcement applications, from IMSI Catcher mobile
phone traffic interceptors (aka "Stingrays") to bulk data collection techniques.[19] The question of whether I am personally subject to what kind
of surveillance would remain extremely subjunctive. As Snowden said,
"you never know."

Meanwhile, government spokespersons deployed subjunctive techniques toward a very different distribution of uncertainties. James Clapper quipped that an electronic surveillance program such as PRISM is no different from fire insurance. But insurance historically developed its legitimacy by quantifying uncertainty into percentages and premiums. The strategic use of disaster statistics and risk percentages could *claim* to provide stable, factual knowledge—despite their often arbitrary beginnings. As we shall see later in this chapter, such calculations are far more difficult to produce with respect to terrorism. In the face of a radically unpredictable threat, state surveillance has practiced, and justified itself on the basis of, a subjunctive reasoning: "everybody"[20] needs to be watched. Within this rationality, surveillance is not, strictly speaking, proved to be necessary by *past* terror attacks or *present* identification of concrete dangers. Proof is always deferred: we must act as if the efficacy of this program has been proven by a danger that, if we are right, we will prevent from ever actualizing.

As If, We Resist

Subjunctive reasoning was widespread not only in the rationalization of and against NSA surveillance but also in the everyday techniques offered to citizens as a way to protect themselves from surveillance. Since the initial leaks, Edward Snowden leveraged his newfound fame to dispense advice on how to stay safe in the dangerous digital world:[21]

MICAH LEE: What are some operational security practices you think everyone should adopt? Just useful stuff for average people.

EDWARD SNOWDEN: . . . The first step that anyone could take is to encrypt their phone calls and their text messages. You can do that through the smartphone app Signal, by Open Whisper Systems. It's free, and you can just download it immediately. And anybody you're talking to now, their communications, if it's intercepted, can't be read by adversaries . . . The other thing there is two-factor authentication.

The term *operational security*, or OPSEC, originates with the U.S. military. The Department of Defense defines it as "the process by which we protect unclassified information that can be used against us."[22] In effect,

Snowden is adding to our already-lengthy list of practices that constitute our "digital hygiene": the things the individual user must do to "kee[p] her data from mixing with others, for avoiding infection with computer viruses, and so forth."[23] Notably, these practices themselves enact a recessive and simulated relationship. Many consumer-end privacy tools provide friendly, easy-to-use interfaces for taking basic steps to alleviate the promiscuity built into our electronic communication networks. It is all too easy: a few clicks, ticking off certain opt-in data gathering services, yellow and white symbols flashing into a reassuring green, and one is allegedly safer. Certainly, there is no doubt that taking advantage of privacy toggles offered by Google or Facebook can limit the commercial exploitation of your personal data. More time-consuming solutions, such as PGP/GPG email encryption, are known to provide some tangible increase in communications privacy. The point is that when privacy tools proudly inform lay users that 42,787 trackers have been blocked so far or that the monetary value of one's personal data to Twitter has declined from $30 to $25, such feedback emphasises the fact that the user cannot meaningfully perceive or keep track of the myriad threats they are exposed to. It is, like the public effort to quantify the Snowden files, a form of painting by numbers, where the proliferation of statistics stands in for the lack of meaningful knowability. This is also the case with examples such as alternative webmail or instant messaging services.[24] The problem is that, as Edward Snowden himself acknowledged, one can never truly become "clean":[25]

> You will still be vulnerable to targeted surveillance. If there is a warrant against you if the NSA is after you they are still going to get you. But mass surveillance that is untargeted and collect-it-all approach you will be much safer [with these basic steps].

Here, "digital hygiene" begins to replicate our modern relationship to bodily health: just as "health" is merely a temporary and relative freedom from pathologies, into which we may slip back at any moment and therefore require our unending vigilance, there can be no escape from the threat of surveillance.[26] And so, like self-surveillance as a technique of care, protecting oneself online becomes an indefinite, routinized ritual. Similarly, the injunction to participate in public speculation—to engage in interminable exegesis of each leak, each public statement—promises some future point at which transparency is achieved but, in practice, enacts a new normal of cynicism. The demand for transparency that follows each scandal, emblematized by the impotent gesture of the online petition, tethers the performances of politics all the more to "what remains to be known." The subjunctive relation therefore entails a *choice to believe*—not the conscious, discrete, sovereign choice emblematized by the good liberal subject but a practical compromise where to participate is to have implicitly chosen to believe.[27] In this way, the essential cast of the fiction is retained: a public that knows for itself, the secret that can/must be revealed, the informed deliberation that can result. To say "we know already" is ultimately a gesture of repair, restoring the sense that the world of knowns we rely on will continue to, by and large, provide solid ground.[28]

In "Publicity's Secret," Jodi Dean argues that the secret is the disavowed basis for the fantasy of the public.[29] She subdivides this (fantasized) public using Jeremy Bentham's taxonomy: the public supposed to know (PSK), a wise tribunal engaged in active knowing, and the public supposed to believe (PSB), a "flux of conflicting opinions" dependent on collective flurries of trust rather than firsthand knowledge. The PSK is the "we" of the "we knew already," while the PSB relies on a subjunctive connection to the knowledge of the PSK. Of course, the entire point is that the PSK retains a certain secrecy from the PSB and that the two maintain a certain difference. This concealment ensures the maintenance of the collective belief that "someone knows" and therefore that "the public" knows. The problem is that in the age of data overproduction and the resulting ecosystem of speculations, the PSB "doesn't believe that the [PSK] knows, and it doesn't need to—mediated technologies materialise this belief as if there were some believing public."[30] The disbelieving PSB takes up the task of "knowing for itself," but far from fulfilling the ideal of the good liberal subject, that public becomes corralled to the incessant feed of new updates, new (mis)information.[31] Insofar as the public is founded on an impossible promise of full and direct knowledge, the very fantasy of the public is dependent on a rotation of things, peoples, and methods by which we believe we know. Subjunctivity enters into this gap as a kind of compensation, an imperfect remedy: a way to convert some of this output of speculation and hypothesis into believable, actionable knowledge.

Interpassivity

If the only people qualified to hold opinions are those who "have all the facts," then politics is not our responsibility. Politics is something that other people do, but not us . . . The public exists elsewhere, not in our town, where regular people live.
—Nina Eliasoph, *Avoiding Politics*

We do not "really" believe that politics may deliver everything it promises [. . . and yet] we do feel an intense connection to the political process and we do expect it to "deliver."
—Gijs van Oenen, "A Machine that Would Go of Itself"

Who—or what—knows for me?[32] Where subjunctivity pulls in potentiality and the future for the work of knowing, such processes often involve the nomination of human and technical others as objects of deferral. It is the NSA's Michael Hayden, assuring the public that he absolutely trusts the NSA analysts with our personal information, and so we should put our trust in the NSA.[33] It is Edward Snowden, or even the files themselves, standing in for large sections of the public that lack access to the full picture. It is the very idea of a general uproar over surveillance, the sense that "the public" or "American society" is working themselves up about it, that justice is being served (or that national security is being compromised). Interpassivity describes a modality of knowing adapted to the distributed and recessive characters of knowledge production in the surveillance society.

Interpassivity originally emerged from art and media theory as a converse to the cliché of interactivity before developing into a theory for a wider range of social processes primarily through the writings of Slavoj Žižek, Robert Pfaller, and Gijs van Oenen.[34] Although each articulation is distinct, they tend to revolve around one general relation: "not me but another for me." Someone else believes, so even if I do not, it remains a kind of "truth."[35] I Xerox a book or tape record a television show and become satisfied that I have nearly consumed it—almost as if the machine has "watched it for me." This subjective "outsourcing" has numerous practical uses. Interpassivity allows subjects to "know" that which may not be supported by their own behavior, identity, and environment in any immediate sense. "I" may not believe in a conspiracy theory, but others do; although I am not offended by a bad joke or violent footage, an unspecified group of "other people" might be. These articulations partially excuse the subject from being bound to the belief in question—even as that belief is hypostatized into assumed reality, forming a concrete basis for opinions and actions. Indeed, "delegating one's beliefs [can make] them stronger than before";[36] claims about states of affairs are externalized from the speaker's own experience, becoming more difficult to dismiss as mere flight of fancy. We are familiar with this mechanism, of course, in the work of rumor. The conceit "I have heard it said elsewhere" holds the truthfulness of the rumor in constant suspense, adding to its resilience. And such interpassive characterizations have been particularly important to mediated ideas of the public. The modern notion of the public has ever been predicated on the ability to imagine silent majorities and other "Others" behind each singular spokesperson or representation.[37]

One prima facie reading of interpassivity is that it is a kind of divestment, a disempowering alienation of the subject from their own activity (including "knowing"). Intersecting with contemporary critiques of participatory politics, Žižek calls interpassivity a form of "false activity" where an upper-middle-class academic may celebrate (or condemn, it matters not) the revolutions of others, or where the VCR will record the television show while the subject labors *as if* rest and relaxation has been had.[38] Yet interpassivity is not only a problem of false activity (which seems inseparable from that old chestnut, false consciousness). To be passive, and to have others do in one's stead, takes a different kind of work on the subject end. If digital activism is a mere simulation of real protest and social media platforms a diluted form of meaningful sociability, the likes, the retweets, and the emails nevertheless constitute a kind of maintenance—the labor required to stay connected to others, intelligible as a public subject, countable as data.[39] Interpassivity cannot be isolated onto a single side of the active/passive coin.

The Humans Supposed to Know

"Here's the rub: the instances where [NSA surveillance] has produced good—has disrupted plots, prevented terrorist attacks, is all classified, that's what's so hard about this."[40] Such was the public defence raised by Dianne Feinstein—at the time, the chairman of the US State Select Committee on Intelligence (also called the Senate Intelligence Committee), a major oversight body for state intelligence activities. This external auditor to the NSA's surveillance programs was now joining the fray to defend intelligence agencies' right to remain shielded from public oversight. The *form* of the apology is a simple one: if only you knew what we know, you would feel as we do—provided, of course, that you believe us that we even know anything.

Of course, such appeals to the secret are so often paired with a controlled leak, an open secret.[41] Even as government personnel insist that the proof of surveillance's efficacy is secret, something about that secret is described, characterized, with impunity (literally, without the ability to be validated). Feinstein was soon joined by Keith Alexander; Mike Rogers, Republican senator and part of the House Intelligence Committee; and, presumably advised by them, President Barack Obama. All three insisted that a very specific number of fifty-four terrorist attacks had been thwarted, "saving real lives"[42] as a result of the surveillance programs in question. A declassified chart, looking much like one of the PowerPoint slides Snowden had leaked, purported to show a breakdown of those fifty-four thwarted attacks into different types of plots and by region (figure 5.1). Yet details and proof remained elusive. Only four out of the alleged fifty-four thwarted attacks have been publicly identified—including the case of Basaaly Moalin, which we will revisit later in this chapter. And for those identified cases, whether bulk electronic surveillance made a difference—that is, whether it was necessary on top of existing tools to prevent the attack—was hard to say.[43] A 2013 report by the President's Review Group on Intelligence and Communications Technologies argued that whereas Section 702–based surveillance of online communications did contribute to many of the fifty-four cases, phone records collected through Section 215 have not made much of a difference.[44] Yet the idea that surveillance "saves lives" continued circulating in public debates. These kinds of claims do more than simply claim the public's ignorance of "all the facts." They also demand that public deliberation take place in full awareness of that ignorance. Just a few weeks after Snowden's initial leaks, Keith Alexander made an appearance at Black Hat, a major information security conference well known for its

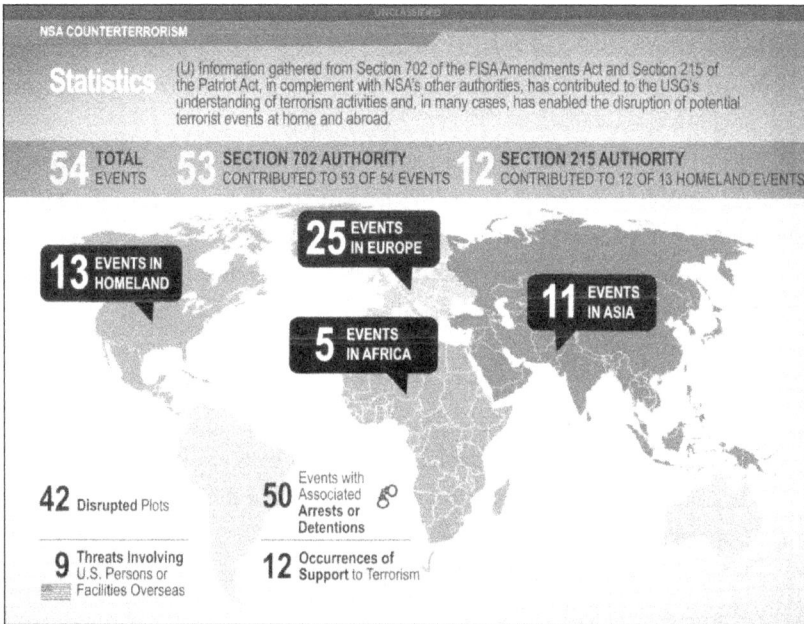

Figure 5.1. The NSA's declassified infographic on the fifty-four plots allegedly foiled through dragnet surveillance, released by *ProPublica*.
Source: Justin Elliott, "NSA 54 Events Chart," DocumentCloud, 2013, https://www.document-cloud.org/documents/802269-untitled0001.html.

hacker contingent. Undeterred by skeptical and angry catcallers in the audience,[45] Alexander pitched his appearance as a no-nonsense presentation of the facts, an occasion for setting the record straight:[46]

> [NSA analysts'] reputation is tarnished because all the facts aren't on the table. But you can help us articulate the facts properly. I will answer every question to the fullest extent possible. And I promise you the truth.

This promise, of course, was couched by repeated reminders that there are things that the audience cannot be told and that some things would have to remain secret. This public relations strategy was not new to the NSA; the organization had already practiced similar rhetoric on its occasional briefings to journalists and other outsiders.[47] Alexander's "facts on the table" often consisted of tautological value judgments. He knows the analysts are great people; he knows that the NSA has great oversight—so would everyone help him spread those facts? In effect, the tech community gathered at Black Hat were being asked to become public advocates for the NSA's truth while being barred from full access to that truth themselves.

A year later, Feinstein's Senate Intelligence Committee held its annual hearing—publicly accessible, and with Director of National Intelligence, James Clapper, present for questioning, as was customary. It was covered widely by the media and streamed online.[48] This hearing—a public affair that is nominally supposed to grant transparent access to the states of affairs and enable democratic deliberation—featured exchanges such as the following:[49]

> SENATOR HEINRICH: Have our European allies ever collected intelligence against U.S. officials or businesspeople or those of other allied nations?
> JAMES CLAPPER: Yes, they have. And I could go into more detail on that in a classified session.
> SENATOR HEINRICH: That's fine, Director Clapper. . . . Do you believe that the Russians have gained access to the documents that Edward Snowden stole, which—obviously, many of which have not been released publicly, fortunately?
> JAMES CLAPPER: I think this might be best left to a classified session, and I don't want to do any—say or do anything that would jeopardize a current investigation.
> SENATOR HEINRICH: That's fine, Director. Thank you, Chairman.
> SENATOR FEINSTEIN: Thank you very much, Senator Heinrich.

The public hearing was an occasion for the public to learn which particular things were not available for the public to know. Such performances asked the reading public to actively hold their judgment in abeyance. More than that, it asked that public knowledge is constructed by simulating the judgment, the affects, of another. To observe such debates and say "See, there is nothing to fear," "The system is broken," or even "We just don't know enough," is to turn to these experience of simulations and deferrals as the basis of judgment in the absence (or, rather, absent presence) of more concrete knowledge.

Not only can the Other know for the subject, but they can also *do* and *experience* in one's stead. Although surveillance's pervasiveness far outstrips the highly infrequent occasions on which it intrudes tangibly into individual lives, interpassivity becomes a key technique by which a given political and affective orientation might be fleshed out into "known" reality. In the wake of the Snowden leaks, commentators often fell back on projected others—ignorant, outraged, afraid, and so on—to populate this verisimilitude. Consider one journalist's anecdote:[50]

> My older, conservative neighbour quickly insisted that collecting this metadata thing she had heard about on Fox was necessary to protect her from all the terrorists out here in suburbia. She then vehemently disagreed that it was okay for President Obama to know whom she called and when, from where to where and for how long, or for him to know who those people called and when, and so forth.

One might read this as a liberal's typically snarky story about a stubborn and misinformed conservative. But the general sentiment that there are people out there, "bad things" happening out there, that need to be watched and stopped is far from an abnormal one. The proliferation of interpassive relations in surveillance discourse enables the neighbor to imagine that somebody else surveilling somebody else is going to personally make her safe. When simulated claims are presented as knowledge, the judgment and action that follow from it also become amenable to such deferral.

As with subjunctivity, the pattern was replicated on the other side of the debate. For many critics of surveillance, Edward Snowden, the heroic truth teller, and his media proxies—primarily Glenn Greenwald and Laura Poitras—formed the authoritative Other who might "know" in their stead. A telling, if probably unintended, gesture to Snowden's interpassive function appears in the final scenes of *Citizenfour*, Poitras's Academy Award–winning documentary about the whistle-blower. Having released his cache of secret files to the world, Snowden has fled to Russia as a fugitive. In the scene, he is visited by Glenn Greenwald, who would like to update Snowden on a new whistle-blower. Possibly as a precaution against eavesdropping, Greenwald scribbles every critical term—the source's name, their means of communication—on a piece of paper, which remains concealed from the camera. Bereft of every key piece of information, the audience is forced to stare dumbly at Greenwald and Snowden as they exchange knowing looks. *The New Yorker's* review, christening Snowden the "Holder of Secrets," was not happy with this choice:[51]

> Several times, Snowden reacts to disclosures that we are not allowed to see; it's as if the viewer were supposed to accept his judgment literally at face value. Poitras has closed a curtain around her main characters, leaving the audience out.

As a criticism of the scene itself, there is not much to say here; the film has a clear moral obligation to protect the identity of this new whistle-blower. But the scene illustrates the relationship that has been established between Snowden the truth teller, the public on whom he wishes transparency, and the information itself. The business of revealing secrets is forced to itself be a secret affair, in exactly the same sense that surveillance must itself operate in the dark while hoarding and trading in the unknown. The secret's incitement to discourse, and the constant need to characterize and operationalize the secret and the unknown, is mirrored across both "sides" of the affair. In fact, within the small community of whistle-blowers on twenty-first-century American state surveillance, Snowden was unusually forthcoming. Consider the State Department official John Napier Tye, who wrote in *The Washington Post* about his concerns over electronic surveillance. Despite often being called a whistle-blower, Tye categorically refused to release classified information. He claimed that[52]

> based in part on classified facts that I am prohibited by law from publishing . . . I believe that Americans should be even more concerned about the collection and storage of their communications under Executive Order 12333 than under Section 215.

Unsurprisingly, the government was happy for Tye to retain his top-secret clearance even after the leaks. His specific criticism of Executive Order 12333, one of the numerous legal provisions undergirding the electronic surveillance programs, was considered part of a very conventional strategy, where government insiders provoke policy debate through controlled but ultimately permissible levels of leakage. Americans were told to be "concerned," but about what exactly, Tye could not say. Meanwhile, legal challenges to state surveillance ran into similar systems of deferral in the juridical sphere. As we saw in chapter 2, lawsuits brought by civil society were stymied by the government's argument that the surveillance practices were too secret to be tried by normal means. As the opinion for *ACLU v. NSA* (2007) explains,

> [D]efendants argue that the state secrets privilege bars Plaintiffs' claims because Plaintiffs cannot establish standing or a prima facie case for any of their claims without the use of state secrets. Further, Defendants argue that they cannot defend this case without revealing state secrets.

The strategy was ultimately successful: the Court of Appeals for the Sixth Circuit ruled that insofar as the American Civil Liberties Union and the persons it represented could not prove they were *personally targeted* by the surveillance program in question, there could be no standing to bring charges. Of course they couldn't; as we saw, most electronic surveillance by the NSA operates on the basis of "selectors," not specific persons, and then ingests vast volumes of electronic communications indiscriminately before determining their candidacy for detailed review. In all this, the American government was following a long-standing tradition on state secrecy. Barry Siegel has shown how as early as *Totten v. United States* in 1875, precedent was set for ruling cases "nonjusticiable" if they would entail the exposure of state secrets.[53] State secrecy privilege has been invoked against numerous other cases against early twenty-first-century state surveillance, from *Hepting v. AT&T* (2006) to *Jewel v. NSA* (2008)—the latter under appeal at the time of writing. The practical impossibility of eliminating recessivity from surveillance produces a pervasive requirement for interpassive compensation.

The Machines Supposed to Know

The other that knows, does, decides, for the subject . . . this other is not limited to human individuals, or even aggregates of individuals, but extends to the machinic. Here we find again the idealization of the machine as a privileged custodian of truth and information. Dianne Feinstein defended dragnet surveillance with an apparently self-explanatory quip: "part of our obligation is keeping Americans safe . . . Human intelligence isn't going to do it."[54] It *was* self-explanatory to a particular community, at least. Feinstein was referring to a basic distinction within the intelligence community between HUMINT (human intelligence) and SIGINT (signals intelligence): between the human business of managing individual sources and the scalable, automatable, machine-dominated collection of electronic communications signals. Thus an NSA strategy statement in 2012, leaked by Edward Snowden, references the present as a "golden age of SIGINT."[55] In such articulations, SIGINT is the inevitable protagonist to the age of the internet and global communications infrastructures. The narrative of SIGINT's necessity thus forms a certain parallel with self-surveillance's vision of automated and autonomous data collection. Both interpellate a world of knowing machines that can produce meaningful knowledge prior to or in excess of human intervention.

Machines that know in our stead loomed even larger in the explicitly fictional and hypothetical. Interpassivity not only does not stop at projecting what is known by others or realized in secret but also leverages more speculative literature to bring forth the secret and uncertain. Although the vast majority of the NSA's SIGINT activities had remained top secret before Snowden, there is also a long tradition of intentional leakages to the public. The CIA is known to have commissioned "strategic fiction" for this purpose;[56] elsewhere, literature from DeLillo to Burroughs and Pynchon and Hollywood films such as *Enemy of the State* (1998) and *Echelon Conspiracy* (2009) have also acted as sites for grasping the concealed reality of surveillance. In other words, these media served as a platform for thinking through surveillance if not as a presently deployed set of practices, then as a set of principles, power relations, and technical paradigms that flit between merely plausible and believably probable. When Snowden first began to leak government secrets, one of the first points of reference for the media and the public was George Orwell. A few months after fleeing Hong Kong, Snowden himself saw fit to reference the writer:[57]

Great Britain's George Orwell warned us of the danger of this kind of information. The types of collection in the book—microphones and video cameras, TVs that watch us—are nothing compared to what we have available today. We have sensors in our pockets that track us everywhere we go.

Yet the comparison was rather redundant. Sales of Orwell's *1984* had already rocketed by some 6,000 percent after his initial leaks in June.[58] The public may not have turned to fiction to take them literally as prophecy, but it could serve as a resource for making sense of the confused present and the uncertain future. Aldous Huxley's *Brave New World* and Philip K. Dick's *The Minority Report* became staple references for communicating the just-unveiled reality of state surveillance.[59] Indeed, in some cases, such writings were explicitly understood by their creators as fiction that combats the deception of the state; that is, fiction is a work of "true lies" that finds its own ground to resist the lies that roam dressed as fact.[60] Meanwhile, state actors also participated in this detour through fiction. In March 2014, the TED Conference invited Snowden and the NSA to speak in succession. The latter's defence of state surveillance included an uncanny reprisal of the technological fantasy in *The Minority Report*:[61]

> If you think about a television murder mystery, they start with the body and work to solve crime. We're starting well before then, before the bodies, to figure out who the people are and what they're trying to do. That involves a massive amount of information.

The machines that know for humans, and machines that know beyond what humans can know: both run amok in this speculative space, fleshing out the skeletal outline featured in more "serious" and prudent discussions of the Snowden files. The narrative of historical shift from HUMINT to SIGINT received its doppelgänger in popular cinema of the 2010s. In *Spectre* (2015), James Bond, that archetypical hero of face-to-face spy work, is threatened not by Soviet-affiliated agents (as he was in *Dr. No* [1962] and *From Russia With Love* [1963]), nuclear weaponry (as in *Thunderball* [1965] and *Octopussy* [1983]), or even nation-specific

organized terrorism (*Casino Royale* [2006]) but an ubiquitous surveillance system coordinated across nine national members of the "Nine
Eyes"—an obvious derivative of the real-life "Five Eyes" alliance that
coordinates signals intelligence (among others) across America, the
United Kingdom, Canada, Australia and New Zealand. Sherlock
Holmes, emblematic of a predigital sensitivity toward physical traces
and other means of human surveillance, was reprised as a contemporary detective in a highly successful BBC series (*Sherlock*, 2010–). There,
Holmes remains the arrogant and mercurial master of HUMINT but is
clearly contrasted by his brother Mycroft, who masterminds state intelligence and functions as a superhuman clearinghouse for a vast web
of SIGINT.[62] And although Holmes may be the hero of that story, it is
made perfectly clear that Mycroft and SIGINT are the domains through
which everything of political and governmental consequence is run—at
least, in this twenty-first-century remake. Even the American superhero
genre joined in with *Captain America: The Winter Soldier* (2014), a film
that consciously worked in a critical depiction of preemptive strikes as
determined by data-mining algorithms.[63]

These fictions provided speculative spaces for public engagement with
realities that often remained too secret and too uncertain to otherwise
attain a firm grasp of. And if the cultural industries were simply reacting
to political controversies in a bid to render themselves topical, they were
also capable of *presaging* the Snowden leaks in the public consciousness.
It has been argued that Hollywood has "softened" the public up for the
shock of twenty-first-century electronic surveillance for years, such that
the cliché "we knew already" references fictional "play" as much as serious investigative journalism.[64] In 2014, *The New Yorker* christened *Person of Interest* as the "TV show that predicted Edward Snowden."[65] The
popular American series presented the public with "the Machine"—an
NSA-style dragnet that "spies on you every hour of every day" and that
the protagonist would use to track down individuals before they became
perpetrators or victims of violent crime. *Person of Interest*'s prophetic
powers had rather mundane roots: the series was conceived through extensive consultation of US state surveillance practices as was known and
estimated at the time.[66] If the Machine is not quite an "accurate" picture
of the present-day capacities of NSA surveillance systems, the vision of
"connecting the dots," of predicting crimes before they happen, and of

profiling individuals based on populational patterns are all played out in *Person of Interest*'s dramatic scenarios. In other words, the technically available information about state surveillance was percolating as much through fiction as news—a circulatory context that favors the fermentation of "half-legitimate" epistemic techniques such as interpassivity and subjunctivity. The television knows for us, too.

It is worth repeating that interpassivity is not simply a form of "fake" knowledge in which the subject pretends to what is not theirs (or are themselves fooled into it).[67] When we fake it, when we pretend, when we bend the rules of what counts as knowledge, that is precisely the daily maintenance work we do as members of the public to sustain a grand fiction—that knowledge is possible, that a rational public is possible. Although interpassivity can and does lead the public to debilitating forms of stasis, to pass it off as an abnormal disease in our knowing underplays its widespread utility.

Zero Tolerance

In Feinstein's apology, or the invocation of state secrets privilege, we find a certain separation: although the public must remain distanced from the truth, this is not the case with the insiders fighting terrorism on the front lines. But as we saw with the lone wolf, the fantasy of data as better knowledge is reprised within counterterrorism operations themselves. The thresholds separating suspicion from guilt, estimate from fact, speculation from proof were modified and improvised as security professionals struggled to reconcile the kinds of knowledge they could produce with the kinds of certainty they had to provide. Professional in their commitment to the virtues of security, counterterrorism officials engaged in gradual shifts in personal judgment and conventions of practice that would enact such modifications.[68] The sloganization of "zero tolerance"—a declaration of war on uncertainty and risk—would demonstrate most clearly the dangerous gap between what is "known" and the increasing demands made of that knowledge.

One key aspect of subjunctivity in counterterrorism has already been subject to extensive research and commentary: the growing importance of simulationionist and otherwise "hypothetical" analysis for the purposes of anticipatory and preemptive interventions. From simulations of terrorist networks based on computational models[69] to elaborately designed, physically enacted scenarios training civilians to respond to terror attacks,[70] the United States and numerous other states have expanded old techniques and developed new ones to actualize and "play with" the possibility of every kind of terrorist attack. These strategies sometimes directly enrolled the public, enjoining them to internalize and practice a form of "lateral surveillance" that some have described as banalization of insecurity.[71] At other times, the state's simulations of dangers might intersect with the media and public imagination of dangers and of state simulations themselves, fertilizing subjunctive grounds for debate and sentiment. In 2015, a map used in the US military's "Jade Helm" training exercises was leaked to the public, provoking speculation and paranoia. The map appeared to show Texas as a "hostile state," which "permissive" states such as California and Colorado might help subdue (figure 5.2).[72] At the same time, a different rumour made the rounds that Daesh terrorist bases had been discovered in Texas. Senator Ted Cruz, "fresh from his Jade Helm inquiry" and just months before he would run for president on promises of getting tough on Islamic terrorists, accused the Obama government of failing to connect the dots. A mock military scenario and an unconfirmed rumour had mutually reinforced each other's status as half-truth or, rather, as operationalizable fiction. In a poll taken immediately after the leaks, 32 percent of registered Republican voters said "that the Government is trying to take over Texas."[73] Counterterrorist logic in early twenty-first-century America features a tight linkage between the virtues of "preparedness" and hypothetical epistemologies.[74]

This section focuses on one specific practice that intersects electronic surveillance and on-the-ground counterterrorist operations: the increasingly systematic practice of producing what sufficiently "counts" as evidence in counterterrorism operations. Here, electronic surveillance collaborates with traditional human intelligence work to coax suspicion into something that can legally count as "real enough." Such strategies have come to occupy a central role in American counterterrorist operations since September 2011—a shift that was accompanied by a belief that the uncertainties surrounding lone-wolves-to-be had to be preemptively dispelled and by a contemporaneous establishment of online communications infrastructures and large scale databases for informants. These

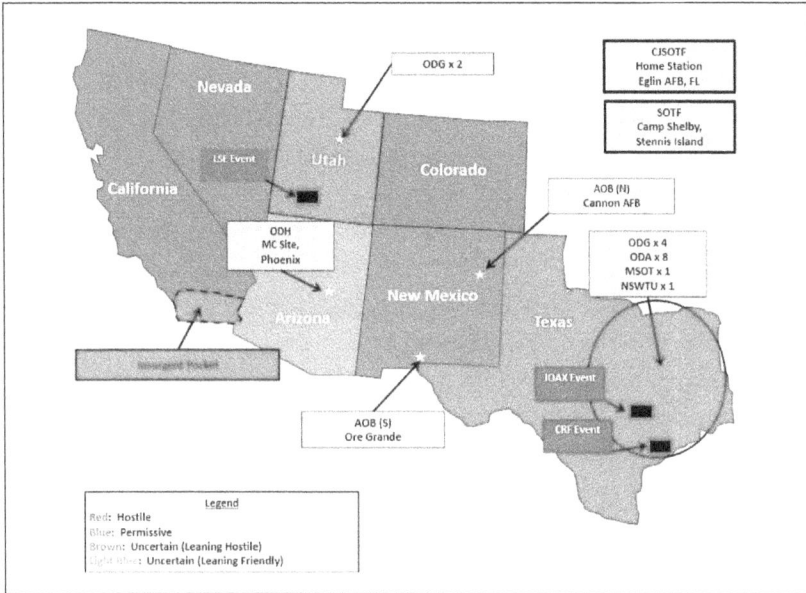

Figure 5.2. A US Army Special Operations Command map of Jade Helm exercises.
Source: Dan Lamothe, "Why Operation Jade Helm 15 Is Freaking out the Internet—and Why It Shouldn't Be," *The Washington Post*, March 31, 2015, https://www.washingtonpost.com/news/ checkpoint/wp/2015/03/31/why-the-new-special-operations-exercise-freaking-out-the -internet-is-no-big-deal/.

are, perhaps, the most obvious sort of fabrications in the book, although not because the individuals prosecuted this way are surely innocent; ultimately, their preemptive arrest means we will never know for sure. As counterterrorist operations grapple with inexhaustible unknowns while attempting to establish sufficient basis for knowledge and action, preemptive fabrication emerges as a "realistic" and risk-conscious practice appropriate to a paranoid epistemology.

Hollywood Endings

In 2011, the FBI began to construct its enclosure of suspicions, records, and proto-predictions around a man named Sami Osmakac. A Kosovo-born American and a Muslim, Osmakac was introduced by a friend to a fellow named Dabus—an FBI informant who, in turn, connected him to an undercover agent named "Amir Jones." To that point, Osmakac's record of suspicious activity included a tendency to verbally criticize democracy, argue for his religion in combative and fundamentalist terms—and a recent arrest for streetside fisticuffs with a Christian street preacher. Something to chew on, little to convict. After meeting Dabus and Jones, however, the situation changed. Osmakac was supplied with money, with which he could purchase fake weapons and explosives prepared by the FBI; he was trained in their use; and he was even given money for a taxi so he could show up to his own attack spot, where he was finally arrested by the FBI. During the process, the FBI agents spoke of Osmakac as a "retarded fool" who needed the agency's support to turn his "pipe-dream scenario" into any semblance of a real threat—a result that they referred to as a "Hollywood ending."[75] The FBI provided material and psychological encouragement that allowed Osmakac to become "dangerous enough" to be legally and operationally eligible for arrest. Of course, this also means that ever confirming whether Osmakac would have acted without such encouragement becomes impossible; the price of a preemptive certainty is the absolute unconfirmability of justice.

Osmakac joined other (predominantly, but not always, Muslim and/ or of Arab descent) Americans whose antisocial, delusional, or otherwise mentally nonnormative conditions were relentlessly manufactured into apparent proof of violent intentions and the capacity to carry them through. Matthew Llaneza was an American of Filipino, Anglo-Irish, and Hispanic descent who was later described by an FBI informant as having "the mind of a little child." Nevertheless, his drunk ravings of "Allah Akbar" at a house party, combined with wild boasts about knowing how to build guns, were treated very seriously in assembling charges against him—overwhelming his family's protestations that Llaneza, with a history of mental illness, was barely capable of raking the backyard as instructed. Eventually, Llaneza too was contacted by an undercover agent, encouraged to plot in words a bombing attack, and duly arrested.[76] Perhaps most famously, the "Newburgh Four"—again Muslims of American citizenship (save one)—were enabled, encouraged, and arrested by the FBI.[77] One report estimates that about 30 percent of counterterrorism convictions between 2002 and 2011 were fabricated through stings.[78] In the case of José Padilla, aka Abdullah al-Muhajir, this subjunctive reasoning was sufficient to approve three years of detention without formal charges—and, possibly, torture (a claim made by Padilla and his lawyers and denied by the government). Paul Wolfowitz, the US deputy secretary of defense, made it clear that the state stands behind this new standard in preemptive action:[79]

> There was not an actual plan. We stopped this man in the initial planning stages, but it does underscore the continuing importance of focusing particularly on those people who may be pursuing chemical, or biological or radiological or nuclear weapons. This is one such individual.

Although the full dataset remains publicly unavailable, *The Intercept* has compiled a list of some three hundred individuals charged with terrorism-related offenses on the basis of such covert operations.[80] Supposed weak indicators predictive of "lone wolf" terrorists, from radical Islam to mental illness to antisocial behavior, reappear over and over again—although many of these individuals tended to work with one or two accomplices, sometimes in international networks. The role of FBI informants and undercover agents are often crucial in making these individuals eligible for arrest. Nelash Mohamed Das was provided with a gun and the fake address of a target (a member of the US military) by an informant—and duly arrested with murder-related charges at the scene. Edward Schimenti and Joseph D. Jones had advocated Daesh causes on social media; Jones, an affidavit for the court case shows, wrote a Google+ essay titled "Jihad: The Forgotten Obligation."[81] Yet it was again a lengthy chain of interactions with informants and undercover agents through which they crystallized a concrete plan. The FBI informant told Schimenti and Jones that he wished to travel to Syria to fight for Daesh; in response, they helped him procure mobile phones and drove him to the airport—concrete forms of action eligible for conviction. Retrospectively, it seems clear enough that the majority of these individuals at least entertained thoughts of violence. Schimenti's last words to the informant at the airport were "[D]rench that land with they, they [*sic*] blood."[82] What is equally clear is that this form of terrorism prevention manages a wider bandwidth of futures beyond the "imminent attack" and that it is, like self-surveillance or digital hygiene, indefinite: there can be no point of cessation, of sufficient security.

Fighting Words

The subjunctive reason at work in monitoring and capture were consistently extended to the juridico-legal sphere. Basaaly Saeed Moalin, a Somali American arrested in 2013, became one of the select few suspects to achieve national prominence. With Edward Snowden beginning his leaks that year, Moalin became citable as an exemplary case for publicly proving the effectiveness of state surveillance. As of 2017, he would remain one of the only cases where the state has publicly cited "critical" reliance on NSA's electronic surveillance. Yet the construction of this claim to certainty was also predicated on preemptive fabrications. Arrested on charges of conspiracy and material support for terrorism—specifically, posting 8,500 USD to a Somali contact associated with the jihadist group al-Shabaab—the prosecution argued that Moalin's frequent phone calls and money transfers amounted to a clear support of terrorism. Once again, Moalin had to be apprehended *before* he could produce any further certainty; it was argued that to wait for "proper" proof would be unacceptably negligent of real dangers posed by the suspect.

In court, the defence directly contested this interpretation of available evidence—and, in doing so, publicly exposed the fabrications as a set of uncertain and primordial indices *oriented* toward certainty. Picking apart Moalin's phone calls collected by telecommunications surveillance, the defense argued that his comments about "jihad" referred to a local jihad in his native Somalia against the Ethiopians, that his money transfers to his homeland had gone to projects for schools and orphanages, and, indeed, that no record showed any definitive statement in support of terrorist attack.[83] The defense went as far to submit to the court alternative translations of Moalin's Somali calls, enlisting cultural interpretations of the Somali material. Moalin's cousins argued that his talk was a well-recognized form of *fadhi ku dirir* (literally "sitting and fighting"), an aggressive but ultimately noncontroversial form of argumentation common among Somali men. Because Moalin was apprehended before he could supply further certainty in the form of a violent attack or concrete statements referring to one, surveillance and arrest had to be justified through subjunctive and paranoid readings of relatively cryptic comments:[84]

MOALIN: We are not less worthy than the guys fighting.

ISSA [acquaintance]: Yes, that's it. It's said that it takes an equal effort to make a knife; whether one makes the handle part, hammers the iron, or bakes it in the fire.

The palpable gap between Moalin's words and the eventual charges (of conspiracy and material support for terrorism) echoes the recessive relationships in the Snowden affair. Although some degree of fabrication is, by definition, a necessary part of any preemptive measure, the 2010s saw a visible embrace of more speculative forms of knowledge that could justify early intervention—largely because it was thought that the evolving state of terrorism did not permit the luxury of waiting for certainty. If these suspects were being directed and shaped on the basis of potential rather than actual danger, operatives and politicians argued, so be it: such preemption is the only way to ever "know enough" in time to stop the next attack.[85] In chapter 3, we saw how the lone wolf as a category acts as a figuration of unpredictability, one that nevertheless requires strategies of approximation and appropriation. These epistemic practices reach their destination point with cases such as Osmakac and Moalin, where new strategies are devised to lift suspicious signs from ambiguity to "sufficient certainty." This subjunctive factmaking does not disappear, either, with more interrogations or stricter incarceration. Rebecca Lemov's analysis of Guantanamo demonstrates the irony: the more that becomes known about these inmates, the more reasons there are to speculate about the danger they pose. Subject to extreme security and indefinite detention, many of these individuals are forced into a judicial limbo—that space of exception in a securitized state where questions of justice remain suspended.[86]

Broken Windows

In 2005, Ehsanul "Shifa" Sadequee was nineteen years old. He was arrested and sentenced to seventeen years in prison for suspicious activity that included translating jihad-related texts, talking to (already-identified) dangerous individuals online and producing a ludicrously amateur "casing video" in Washington, D.C., with an acquaintance named Syed Haris Ahmed (who would also be arrested). The low-quality footage, shot by Sadequee, swivels wildly, flashing locations such as the Capitol; Ahmed, in charge of driving, can be heard shouting, "Take it, take it! Allah Akbar . . ." Later, another voice—presumably Sadequee—narrates that "this is where our brothers attacked the Pentagon."[87] In Sadequee's case, the role of informants as a technique for coaxing evidence, if any, is unclear; no publicly available details indicate such. In a period during which behavior such as Osmakac's was sufficient for arrest, Sadequee had made no direct move for a violent attack but had clearly done enough.

Years later, filmed by a documentary crew, a rare kind of meeting was held: Sadequee's family, on a protracted quest to prove his innocence, journeyed to meet Philip Mudd. As deputy director of the National Counterterrorism Center at the time, Mudd had had a direct hand in the case and the decision to arrest. Although courteous and sympathetic to Sadequee's family, Mudd insisted on the necessity of such an action:[88]

> People like me are in a difficult position. We cannot afford to let dozens of innocent people die because a youth makes a mistake . . . If we switched roles, what would you do? What would you do? Would you let him go?

Mudd's dilemma was a classic one, intensified by the political and moral climate of the times.[89] Counterterrorism experts spoke of a "zero tolerance" climate permeating the agencies in the years following September 11; the idea that even a single further terrorist attack on American soil was absolutely intolerable and had to be stopped at all costs. Andrew Liepman, who had been director of the National Counterterrorism Center from 2005 to 2012, described the political climate in that period as one of "zero failure, zero attack threshold."[90] Such a climate would intensify after every attack that nevertheless occurred. Across the Atlantic, a series of major suicide attacks had become a central issue in the French presidential elections of 2017. The eventual victor, Emmanuel Macron, insisted: "I propose pragmatism, with zero tolerance"—with terrorism, with crime, with delinquency.[91] His rival, the far-right Marine Le Pen, was not to be outdone: "in the face of gang violence, the state must be unyielding and respond with zero tolerance" to overcome an existing "culture of permissiveness."[92] A month later, three Muslim individuals attacked London Bridge and

Borough Market, killing eight. The British Prime Minister Theresa May announced—to a city famous for its massive population of CCTVs— that "there is, to be frank, far too much tolerance of extremism in our country."[93] What does it mean to be "tolerant" of extremism or terrorist attacks? Zero tolerance is an impossible and oxymoronic concept, a piece of politically overcharged rhetoric designed to soothe the trauma of a fresh terrorist attack. Decision-making within actual counterterrorism operations are far more cognizant of the irreducible uncertainty in the predictive process. Nevertheless, Mudd's words point to the ways in which this political climate moves the operational needle. The multiple strategies of preemptive fabrication around individuals such as Sadequee were necessitated by the sense that what used to count as due process, as sufficient thresholds of certainty, were too slow, too dangerous. The preemptive actions taken might themselves carry a margin of error, but—like the NSA's "incidental" collection of domestic communications data—articulated as necessary. One form of uncertainty was being traded for another.

The concept of zero tolerance had long lurked in the background of American discourse on crime and discipline. The term itself was born in the American drug policy of the 1980s and, during the 1990s, was gradually adopted for the codification of aggressive enforcement for law and order across schools and local police.[94] It is consistently associated with the notorious "broken windows" theory, which provided a general framework for linking every small element of disorder with a wider onset of serious crime and violence.[95] Rudy Giuliani and William Bratton, the mayor and police chief of New York, respectively, conjured images of a city "out of control" that had to be matched by an aggressive and comprehensive cleanup.[96] This legacy is reprised in the contemporary depiction of a terrorism out of control and escalating aggression as its only deterrent. In the context of state surveillance and counterterrorism, zero tolerance renders epistemic uncertainty intolerable. It becomes far more difficult to respect the rights of suspects because one cannot write off any attack as an "acceptable" or unavoidable loss. Yet, in so many cases, especially that of lone wolves and "homegrown" terrorists, the possibility of crime remains uncertain until it is too late to intervene. These gaps would be filled by a combination of more "actively" engaged informants, subtle shifts in the judgments of experts such as Mudd, the degree of aggressiveness in pursuing court rulings, and other gradual shifts in operational standards of what counts as sufficient proof.

The imagined end point of zero tolerance takes us back to the fantasy of epistemic purity—the idea that the lone wolf, the terrorist, the delinquent can be identified and excised from the social body with the help of advanced technologies. Even as the actual implementation of surveillance involves a complicated redistribution of uncertainties, the publicized fantasy of prediction and control is wielded as a flexible solution for many different fears of impurity and contagion. One of the consequences is a certain overconfidence in the data at hand, pushing institutions to squeeze maximum action out of available knowledge. It is a strange and unsettling equation: the more necessary the action, the less we shall question the data.

Virtuous War

The changing standards of truth and proof in counterterrorism were also consistent with the wider political and military climate. The search for certainty in a world out of control provoked numerous other strategies by which a clean cut might be achieved between citizens and monsters, secure and unsecured zones, war and peace. In this sense, the problem of producing sufficient knowledge out of Moalin's communications is formally analogous to the most infamous case of speculative proof in the period: the alleged presence of weapons of mass destruction [WMD] in Iraq, which functioned as a crucial casus belli domestically and internationally for American invasion. In February 2003, Secretary of State Colin Powell famously presented the US administration's case for war to the UN Security Council. Among the "facts and conclusions based on solid intelligence" he offered were foreign communications data intercepted by the NSA. In the first excerpt, an Iraqi colonel vaguely referred to "this modified vehicle . . . what do we say if one of them sees it?" In the second and third, there were equally broad references to "forbidden ammo"—and then, in a rare moment of relative specificity, an order to "remove the expression 'nerve agents' from wireless instructions."[97] From arresting Muslim American teens to justifying regional war, the accepted standard of what constitutes "sufficient" certainty was adjusted to match an allegedly more uncertain reality.

Another contemporaneous development was the emergence of drone warfare as a major player in military operations. The Obama administration enthusiastically embraced unmanned aerial vehicles (UAV), and a doctrine of "clean" warfare that would take the fantasy of epistemic purity to the battlefield. High-tech surveillance and its predictive knowledge production would erase enemy combatants—and only enemy combatants—before the messy and contagious violence of traditional warfare could begin.[98] Of course, no technology is guaranteed an infallible claim to effectiveness but must socially construct standards and measures such as accuracy and precision.[99] Since the first confirmed use of UAVs in a "kill operation," variously pegged as late 2001 or early 2002,[100] drone warfare has been dogged by public consternation over how this complex, human–nonhuman system "knows" to kill.[101] Paralleling the state defense of NSA surveillance, the justification of drone strikes have often fallen back on secret proof and publicly unverifiable knowledge.

What we do know about these fabrications tend to parallel the aggressively justificatory strategies on the counterterrorism side. Snowden-leaked files on drone operations in the Hindu Kush between May and September 2012 show that unknown casualties tend to default into the category of "enemies killed in action" (EKIA). The victims of drone strikes, eviscerated from afar, thus "count" toward the justice of their own deaths until proven otherwise. Ultimately, the said operations—dubbed HAYMAKER—reported 54 drone strikes yielding 157 EKIA but only 19 "jackpots," or specific targets confirmed as dangerous individuals.[102] Often, the victims' basic demographic identity as "military-age males [MAM] in a strike zone" was sufficient to mark them for a speculative and deindividualized death.[103] Visual cues such as praying or holding "cylindrical objects" also supplied indicators for licensing drone strike in ways that remind us of the use of "weak indicators" in the search for lone wolves.[104] Speaking about the case of Basaaly Moalin, the NSA's Keith Alexander explained the logic at work: to find the bad guys, "I need to know who his network of friends are, because chances are many of them are bad, too."[105] Where Arab Muslims in the Middle East were fabricated into targets—and then killed—based on physical proximity, Arab Muslims in the United States were fabricated into threats and arrested based on communicative proximity. Across the knowledge production work in drone warfare and sting operations, we find the same problem: the ethical consequences of a system of classification and a standard for sufficient proof, one whose choice of what kinds of proximity and association to privilege is often passed off as "common sense." The proliferation of subjunctive reason legitimizes an environment where being "close enough," physically and metaphorically, is increasingly enough to count as guilty.

This compression of the distance to guilt, of the distance separating suspicion and proof, is compounded by another kind of distortion. Drone killing reflects the shrinking space-time of decision in automated, distributed systems. On one hand, there is the overwhelming pressure to act, to save the right lives and take away the right ones, to not miss the smallest window of opportunity. On the other hand, the decision—which, make no mistake, must still be made and still is open to human judgment—becomes strung out across pilots, commanders, analysts, and the machines themselves.[106] The relay of information and judgment across these networked systems typically involves an average of thirty to forty-five hectic, real-time deliberations; this entire process is envisioned to quicken into "seconds" by 2025.[107] Such compression does not eliminate the need to render judgment over uncertain problems, but it does change the wiggle room human subjects have to steer the process. Although covert operations such as those on Sadequee or Osmakac often take months and years to develop, they exhibit similar pressures between the injunction to act and an urgency compelled by uncertain timeframes (you never know when your suspect will "snap" and make an attack). This aspect of counterterrorist fabrications—the structure of decision vis-à-vis uncertain realities and the risks of action—we turn next.

Zero-Degree Risk

September 11 cast a long shadow of terrorism over America in the early twenty-first century. And yet, some voices cautioned that all this might have been blown a little out of proportion. After all, the statistical probability of death from a terrorist attack in the United States between 2007 and 2011 was about one in twenty million.[108] On numerous occasions, President Obama took to reminding both the public and his own staff that "the odds of people dying in a terrorist attack, obviously, are still a lot lower than in a car accident"[109] or even falls in bathtubs.[110] *The Washington Post* pointed out that "you're more likely to be fatally crushed by furniture than killed by a terrorist."[111] Yet the invocation of such statistics (which hardly dampened the dominant discourse of fear) illustrates the troubled relationship between surveillance, terrorism, and probabilistic calculation—and the centrality of the latter in producing working standards of "sufficient" knowability. On one hand, what knowledge can statistics deliver about an event such as September 11, a devastation of a singular significance in the context of American history, or a figure like the lone wolf terrorist, a figure that hides in the many residual interstices between known demographics and modes of organization? On the other hand, the play of correlations and probabilities was precisely how data-driven surveillance promised to render such events predictable and preventable—and, even, how the benefits and harms of such surveillance are assessed. The impossibility of statistical risk calculation was intertwined with a renewed vision of total calculability.

This tension consistently broke out into public discourse over the course of the Snowden affair. The question of terrorism's probability—and the calculation of surveillance's effectiveness—became a key site for claiming and contesting the legitimacy of surveillance practices. An overarching frame was furnished by the frequently expressed notion that the problem of surveillance is to find the right "balance," to strike the right "bargain," between the conflicting values of security and privacy. Soon after September 11, in a statement to the US Congress (made available to the public), NSA director Michael Hayden insisted:[112]

> What I really need you [Congress] to do is talk to your constituents and find out where the American people want that line between security and liberty to be . . . We need to get it right. We have to find the right balance between protecting our security and protecting our liberty. If we fail in this effort by drawing the line in the wrong place . . . the terrorists win and liberty loses in either case.

He would repeat the idea of "balance" several times—in his 2006 address to the National Press Club as a Deputy Director of National Intelligence, and later in his autobiography as a retired general.[113] Barack Obama's public defense of surveillance programs invoked a similar bargain:[114]

But I think it's important for everybody to understand, and I think the American people understand, that there are some trade-offs involved . . . And the modest encroachments on privacy that are involved in getting phone numbers or duration without a name attached and not looking at content—that on, you know, net, it was worth us doing.

This idea of a grand "bargain" asked the public to weigh the pros and cons of state surveillance and to balance the equation toward a Goldilocks point where harms might be minimalized and benefits maximized. It was a classic case of risk-oriented, actuarial thinking. Yet efforts to deliberate toward this balance exposed a deeper epistemic problem: How could surveillance be "balanced" if its effects cannot be assessed in meaningfully quantifiable terms? What kinds of statistical reasoning and risk calculus could be enacted when it involves the collection of literally uncountable pieces of data with the aim of preventing even a single attack (under the full awareness that absolute security is impossible)? Is a mathematical balancing of the various moral obligations vis-à-vis surveillance even possible? Consider contemporary efforts to ascertain the value of surveillance. *The New Yorker* asked:[115]

the N.S.A. has [collected] records from hundreds of billions of domestic phone calls . . . the government has not shown any instance besides Moalin's in which the law's metadata provision has directly led to a conviction in a terrorism case. Is it worth it?

This tension between "zero tolerance" and the probabilistic nature of modern risk can be found across other counterterrorist efforts. Harvey Molotch describes a multimillion-dollar surveillance system installed in New York's subways during the mid-2000s, a project in which the political demands for concrete action overcame the transport authority's own skepticism. As of 2012, the system had yielded a small number of arrests for misdemeanors and not a single lead related to terrorism; in other words, its "cost" in the form of money and civil rights had yielded no tangible "benefits." Meanwhile, actors continued to manipulate numbers and statistics as a way to shape the epistemic field: "1944 New Yorkers saw something and said something," insisted a poster produced by the transport authority, although the number has yet to be traced to any actual documentation.[116] Yet the point of a subjunctive knowledge environment is that the absence of proof does not authorize certainty one way or the other; after all, unknown terrorists may have been deterred from attempting terrorist attacks by the presence of cameras. Or the system may yet live to catch a terrorist red-handed—if not tomorrow, then next month, next year . . . As we saw with counterterrorist strategies for coaxing suspicions into proof, terrorism had become an epistemic problem not so much concerned with minimizing probabilities and constructing acceptable models of risk, but a politically charged search for absolute certainty in a context where a degree of unknowability is inherent. The unfalsifiability of surveillance's value proposition reflects the difficulty, or perhaps impossibility, of producing a morally sound "balance" between security and civil rights.

These debates around terrorism's probability and surveillance's efficacy reflect an ongoing struggle to rationalize these uncertainties in terms of risk and, specifically, to derive probabilistic calculations by which the costs of terrorism or surveillance might be defined (and therefore justified, mitigated, and compensated). In other words, the appeal of risk was that uncertainty could be "frozen" into a usable and reliable bandwidth of possibilities. In James Clapper's invocation of fire insurance, the language of risk functions as a skeuomorph, translating accumulated familiarity and legitimacy to new efforts for conceptualizing the dangers of terrorism (or surveillance itself). Yet uncertainty has never been entirely reducible to risk. The exemplary distinction was already provided by the economist Frank Knight in the 1960s, when he differentiated measurable, and therefore calculable, degrees of chance (risk) from the kinds of unknowns that cannot be accounted for (uncertainty).[117] This did not mean relegating uncertainty to a formless mush that should be bracketed out of serious predictions. Rather, Knight argued, this uncertainty faced by each economic actor produces asymmetries, zones of ignorance, and suboptimal decisions, wherein change and profit might be generated. In anthropology, a group of scholars have examined the sites of actual operationalization where strategies for managing uncertainty go beyond the rigidly calculative and rationalized.[118] Elsewhere, we find terms such as *agnotology*—the study of ignorance—to address practices of deliberately producing and perpetuating uncertainties, the exemplary case being the work of tobacco lobbyists.[119]

Risk is not an objective expression of uncertainty but a particular technology for converting and reducing them into a transactional object—and one that is increasingly ill suited to the kinds of uncertainties that data-driven surveillance seeks to eradicate.

This disjuncture has not gone unnoticed by theorists of risk.[120] Risk itself is not a singular, static concept, and each localized conception of risk attempts to capture uncertainty in different ways.[121] Ulrich Beck, among the most prominent of risk scholars, had come to fame through *Risk Society*, in which the modern attitude to risk defined by a rigorous economy of "acceptable levels" (i.e., acceptable failures, deaths, disasters) that imposes a clear grid for calculating optimal decisions and practices.[122] By 2009, near the end of his life, Beck had turned to the argument that societies are experiencing a swell of unknowns and unknowables and that responses to climate change and terrorism characterize a "planetary state of exception" to the rule of risk calculability.[123] However, as Mitchell Dean has argued, establishing a reliable standard for arguing that the world has become "more dangerous" in realist terms is difficult.[124] The point is that risk has always been a question of how dangers are perceived and calculated by societies—and in this epistemic sense, the differences must be counted.

It is worth remembering that the "traditional" conception of risk had itself emerged out of a proliferation of data and widespread enthusiasm about its epistemic capacity. As new kinds of statistical information became available and "the world teemed with frequencies,"[125] scholars and experts of the nineteenth century rushed to establish laws and models that could leverage this data toward new knowledge, more certainty. This heyday of modern risk included both the expansion of actuarial calculations (for, say, sickness and mortality) and ultimately unsubstantiated efforts to identify "criminal" faces or "inferior" skull shapes.[126] Such rationalizing work often swept away (although not completely) many older heuristics for danger, luck, and chance, but this conquest was itself achieved through a host of often arbitrary decisions. Ian Hacking relates a striking example in the 1820s: the British Parliament, eager to establish actuarial laws to facilitate the fledgling business of life insurance, summoned one John Finlaison, the first president of the Institute of Actuaries—who proceeded to tell them that no such calculation was possible at the time.[127] Rather than heed his expertise, however, Parliament cajoled Finlaison repeatedly until the man grudgingly produced some numbers. The normalization of risk logic itself involved a set of heuristics and veridical objects for rationalising unknowns into knowns (or, rather, probabilities). Debates around data-driven surveillance continues to invoke those high modern techniques—the language of statistics, probabilities, trade-offs, equations—to mitigate the rising feeling of radical uncertainty.

If each social construction of risk appears to harbor some unquantifiable excess of risk within it, this tension is traceable right down to the birth of the word itself. Catherine Althaus shows that although the exact origin of the modern English term is disputed, *risk*'s various possible roots all derive from a sense of uncertainty beyond quantification and calculative control.[128] In one telling, the lineage stems from the Arabic *risq*, meaning that which has been given by God and through which one profits. Another theory hearkens back to maritime dangers faced by European sailors, from medieval Spanish *risco* (rock) to the Greek *rhiza* (hazards of sailing too near to the cliffs). Modern risk has perennially struggled to capture and eliminate uncertainty, even as it constantly designates new regions in which calculation is considered implausible. State surveillance, of course, is one such insurance against the dangers of post-9/11 terrorism. The insistence that the right balance has been struck—and underlying that, the presumption that such an equation is possible—enacts a social process of fabrication whereby surveillance is able to somehow prove its worth against the dangers of terrorism that is somehow able to be known and affixed with a value. What we see here is no epochal shift where technologies of risk reach a "limit point" but the smuggling of less quantifiable material into the operational theater of risk epistemology.

* * *

Even as data-driven surveillance and its war on terror continue to borrow from the language and authority of older forms of risk, the forms of uncertainty that they conceptualize and operationalize do not fit neatly into those statistical epistemologies. The resulting contradiction might be termed "zero-degree risk": the intersection of a veneer of statistical reasoning with a certain threshold of indeterminacy and negligibility, which defaults to a more speculative and preemptive kind of claim making. It is to say that we have to calculate the risks and act on the numbers, but because the threat we face is so uncertain and mysterious, we must be ready to elevate the data we have into the knowledge we need.

Insofar as terrorist threat is conceptualized as a zero-tolerance event of incalculable harm, it defies simple mathematical calculations of value. If the risk of death from terrorism is lower than from a car crash, how low does it have to be to fall under "acceptable levels"? If a given surveillance program violates the rights of several million citizens to catch a single terrorist (or, as is often more likely, preemptively apprehend a suspect), is it "worth it"? In contrast to financial trading or gambling, where a certain ineliminable degree of uncertainty is accepted as the price for the benefits of risk and probabilistic reasoning, the post-9/11 climate embraced the fantasy of total security and total preemption. When the "what if" side of the equation is as catastrophic as September 11, the "value" that each rights violation holds on the opposite side of the equation does not just decrease: it becomes unquantifiable. Surveillance practices are often justified on the bargain claim that a certain amount of invasion of privacy, for instance, is "worth it" for a certain amount of safety gained—as we saw with Obama's defense of NSA surveillance. This implies lines in the sand, at least, where we could say surveillance *is not* worth it. Yet this is not the case in a "zero tolerance" climate, where there is a constant emphasis on the calamitous enormity of each terrorist attack. When even a single error, a single oversight, could cause apocalyptic harm, surveillance becomes an emergency measure that can never be repealed. In this broken equation, surveillance has neither a proper "fail-state" (where it is proved to be inappropriate) nor a "success-state" (where it can be proved that it has done the job). Its failure is always provisional, and its success can always be positioned as just around the corner—thus fulfilling an analogous function to how the innovations of the "next upcoming version" constantly endow self-surveillance technologies with an air of prospective fulfillment. Surveillance's obsession with potential futures means its calculation of "worth" and "success" increasingly become indifferent to actualized cases of danger.[129]

This is not to say that surveillance and terrorism in the twenty-first century is "post-risk." Actuarial reason might still declare a certain individual to be "worth" a certain monetary premium; the 2002 Terrorism

Risk Insurance Act indeed incentivized insurers to provide packages for terrorism.[130] The effort to classify and predict the lone wolf, as witnessed in chapter 3, demonstrates the enduring attraction of correlational and probabilistic calculations of risk. Zero-degree risk expresses the ways in which surveillance and terrorism—in public debate as much as its internal operations—struggle to establish a calculative reason or, at least, a performance of such reason over the incalculable. The proliferation of zero-degree risk as a heuristic reflects the fact that even as the vision of data-driven predictivity and total archives articulates surveillance's ideal form, both its practitioners and the public are faced with the practical need to make decisions and assessments.

This "upselling" process, where incomplete data and provisional judgments are fabricated into authoritative predictions, also relied on a moralizing discourse: that doing *something* must be better than doing nothing, and therefore, "everything" that can be done must be done. Matthew Hannah calls it "actionism"—an intensification of a propensity for action and the performance of being "proactive" that is latent to modern politics.[131] One oft-cited instance is British Prime Minister Tony Blair's defense of the American invasion of Iraq, itself very much tied to the political fascination with terrorist threats at the time:[132]

> But sit in my seat. Here is the intelligence. Here is the advice. Do you ignore it? But, of course intelligence is precisely that: intelligence. It is not hard fact. It has its limitations. On each occasion the most careful judgement has to be made taking account of everything we know and the best assessment and advice available. But in making that judgement, would you prefer us to act, even if it turns out to be wrong? Or not to act and hope it's OK? And suppose we don't act and the intelligence turns out to be right, how forgiving will people be?

Here, we might recall Philip Mudd's words when he, from the position of a counterterrorism specialist, explained to the families of a fabricated terrorist why such engineering of proof was necessary: what if you understood that your judgment, an act of *faith* either direction, stood between convicting an innocent or allowing mass murder to happen? "What would you do? Would you let him go?" The same question echoes silently in the networked centers of drone strike operations, as well as justifications of measures such as CCTVs in schools.[133] And, of course, the NSA's post–September 11 strategy—to collect "everything" because anything might be important—is itself the clearest manifestation of actionism. The contradictions between an absolutist, zero-tolerance politics and an uncertain world of catastrophic risks are sidestepped through an ethics where one always opts to do, to save, to record, and to arrest, an epistemology where the known harms, or "side effects," of surveillance are counted for less than the unknown harms of inaction.[134]

Actionism completes the zero-degree equation and rationalizes a pathway from uncertain epistemology to concrete decisions. Here, the *possibility* of future harms is prioritized over (or, rather, enters in place of) probability, completing the irony of the mathematical veneer; precisely the things that escape statistical knowability are presented as a firm basis for "knowing enough" to act. Brian Massumi references the "decisionism" of Obama's predecessor, George W. Bush, as an example of this attitude toward decisions and uncertainty.[135] As the sitting president when the September 11 attacks occurred, Bush would pursue a rationality that does not dither and contemplate the specifics of danger because it cannot be fully clarified anyway; what is important, the argument goes, is to *do* something about it with certainty. Today, that "something" defaults to surveillance. Zero-degree risk becomes a crucial element in the self-legitimating loop between risk and surveillance. In this loop, surveillance evacuates itself from external scrutiny that might regulate its standards of sufficient proof (for the guilt of a citizen, for the efficacy of the system, for the danger levels of terrorism). Its developing array of fabrications, tightly plugged into the affects and performances of security, retain the guise of rational and evidence-based reasoning—indeed, the guise of the latest innovations in that area—even as it provides new channels for mobilizing the uncertain and the unknown for action. Karl Rove, one of Bush's key advisors in the period, pointed out,[136]

> We're an empire now, and when we act, we create our own reality. And while you're studying that reality . . . we'll act again, creating other new realities, which you can study too, and that's how things will sort out.

But this is not, as some have called it, a "post-truth politics,"[137] at least not in the sense that such politics does not care about the truth or is willing to throw away the last pretenses to it. The aspiration toward the regulatory ideal, or, at least, its performance, remains as strong as ever. It is rather the development of a certain pragmatism, a rebalancing of what is considered necessary in the face of enduring uncertainties in a data-saturated world. It is a shifting paradigm for what counts as truth and what you can do with that truth, a shift that evolves gradually out of a distributed set of seemingly independent practical dilemmas, whose similarities can be traced to their common reliance on the fantasy of objective data and technological progress. This reliance on fantasy, and its long modern lineage, is where we turn to next.

6

Data-Sense and Non-Sense

All epistemology begins in fear—fear that the world is too labyrinthine to be threaded by reason; fear that the senses are too feeble and the intellect too frail; fear that memory fades, even between adjacent steps of a mathematical demonstration; fear that authority and convention blind; fear that God may keep secrets or demons deceive. Objectivity is a chapter in this history of intellectual fear, of errors anxiously anticipated and precautions taken.
—Lorraine Daston and Peter Galison, *Objectivity*

Fabrications are borne out of renewed ambitions for control through objective truth—ambitions equally founded on the fear of uncertainty.[1] The heuristics for deferred and simulated knowing in counterterrorism sought to reprocess such uncertainties into usable, believable truths, yielding a stable platform for reason and morality. Again and again, imperfect technologies—full of messy data, presumptuous categorizations, unprovable conclusions—are pressed into service by the demands of honeymoon objectivity. In the domain of self-surveillance, the imperfections of datafication are even more clearly visible. Here, half-functioning prototypes and spectacular press conferences constantly engineer precarious, yet effective, overtures toward a posthuman future. These imaginary media promote a posthuman vision of sensory augmentation that I call "data-sense": a gradual merging of human and machinic sensibility, normalizing and naturalizing new channels for knowing the world out there and the body in here. Emerging across the promissory discourses of early adopters and entrepreneurs, serial tech prophets and corporate sponsors, data-sense constitutes an ad hoc theory of how human bodies are being directed to internalize new knowledge regimes of a data-driven society to stay productive, legible, efficient, and optimized.

Data-sense can be understood as a popular narrativization of big data and machinic sensibility as a new sensory default. The rising prominence of self-surveillance technologies, and big data analytics more broadly, provoked heady expectations that human subjects would (and should) soon internalize these new forms of data-driven knowing into a default and preconscious reference point for self-knowledge. This technological dependency was presented as a historical inevitability, an impending transformation of society already baked into the trajectory of technological progress. Thus, consumers are advised to practice the same anticipatory mindset that characterizes state surveillance and to rush to adopt the technologies of tomorrow in order not to be left behind. Converse to the anxieties around the uncertainties of terrorism and the demand that datafication help secure borders and populations through data-driven predictions, the ebullience of data-sense seeks objectivity as a futuristic rehabilitation of individual control and a world of stable meaning. Dreams of objective truth and an ordered, controlled world return in yet another honeymoon period: with self-surveillance, with the latest smart sensors and anticipated inventions just around the corner, we might finally know ourselves objectively.

Yet, as we saw in chapter 4, this promotion of self-surveillance as a path to human knowing enacts a crucial displacement, installing new processes of fabrication that deal in knowledge not by or for human subjects. This new epistemic order troubles enduring conceptual frames for what it means to resist institutional power, assert individual agency, or defend the spaces of human freedom. It further imperils the mythical figure of the good liberal subject—who knows for themselves, who intakes information and produces choices, who opposes control with freedom—that supports so much of our moral and political thinking. Instead of a tidy and linear progress toward better knowledge, self-surveillance thus reprises modernity's long-running tensions between the ideal of knowing for oneself and the increasingly complex and non-human systems through which that knowledge is to be produced. The angel moves toward the future, yet his face is turned toward the past.

Unintelligible Intelligence

When machines get it wrong, how they get it wrong can be revealing. In 2015, Google's image recognition algorithms infamously labeled a black woman with a "gorilla" tag. The algorithm's failure echoed not only the whiteness of many training datasets used to develop such systems but also the fundamental alienness of a perceptual system that produces knowledge based on indifferent correlations of visual data broken down into meaningless fragments. This incommensurability[2] marks a critical shift in machinic sensibility—one that we may, with Friedrich Kittler, trace back to the later nineteenth century.[3] Chapter 4 described the ways in which machinic sensibility and the knowledge it produces goes beyond the capacity of the human senses. Not only does a tracking device sense the body in ways that the conscious subject cannot (for instance, their galvanic skin response), but machine learning also yields predictions and associations that neither the algorithm nor its human creator can explain in (humanly) comprehensible terms. Here, the problem of recessivity reappears as the problem of unintelligible intelligence: as one *Wired* piece noted for its readers, "our machines now have knowledge we'll never understand."[4]

In *Cloud Face*, the artist duo Shinseungback Kimyonghun (SSBKYH) displays images of clouds that such algorithms misrecognize as human faces. In them, the human observer catches fleeting glimpses of what the machine must have seen—echoing a millennia-long history of humans seeing shapes in the sky (figure 6.1b). Yet such moments emphasize differences more than similarities. In *Flower*, SSBKYH present images of heavily distorted flowers that the algorithm nevertheless identifies "correctly" (figure 6.1a). Although these flowers typically do remain recognizable as flowers for human observers, they represent a growing divergence between the flowers that count for machines and the flowers that count for us. What, in the end, is the sense of a flower, a face?

The overlapping moments in these alien sensibilities are constantly exploited for new kinds of fabrications. As Google was wrestling to refine its algorithms, a new kind of machine-produced, alternative sense broke into the spotlight. In late 2017, a free program named FakeApp brought recent advances in machinic image recognition and artificial neural networks to the public. FakeApp allowed users with modest technological literacy and basic computer equipment to merge foreign images onto video clips with relative ease, speed, and sophistication. These videos, enabled through recent advances in deep learning techniques, would earn the general label of "deepfakes." Predictably, the first popu-

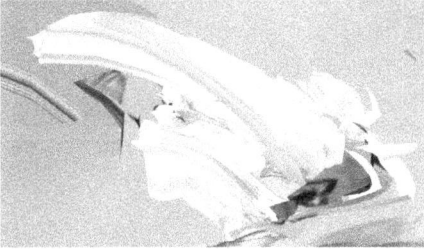

Figures 6.1a and b. In *Flower* (left), Google's Cloud Vision API is used to identify images that the system reports as a flower with high confidence. In *Cloud Face* (right), the artist duo Shinseungback Kimyonghun display images of clouds that such algorithms misrecognize as human faces.
Source: Seung back Shin, and Yong hun Kim, *Flower*, Shinseungback Kimyonghun, 2017, http://ssbkyh.com/works/flower/; Seung back Shin, and Yong hun Kim, *Cloud Face*, Shinseungback Kimyonghun, 2017, http://ssbkyh.com/works/cloud_face.

lar applications were almost entirely pornographic, quickly spawning a sizable archive of artificial videos for human consumption. The fact that these videos could titillate and/or disgust human subjects (at least, enough to motivate their continued production) spoke to their effectiveness; here was a kind of machine-produced nonsense, now used to disrupt and manipulate traditional heuristics of human sensibility. Sexual desire proved amenable to machinic excitation as well as machinic tracking.

The advent of deepfakes, and the crisis it caused in the veridical credibility of moving images, extended earlier dilemmas that concerned static images.[5] Smaller in file size, and easier to edit and circulate, photographs and other nonmoving images had already begun the path to alien unintelligibility (or, rather, alternative intelligibility). Trevor Paglen observes that "the overwhelming majority of images are now made by machines for other machines, with humans rarely in the loop."[6] Some of these images, such as SSBKYH's flowers, bear some material traces of human signification—although their primary channel of legibility lies with machines. Even the common lament that there are *too many* images indicates a shift in the interaction of human and machinic sensibilities. It is not only that certain kinds of data become unreadable to humans, locking them out of forms of meaning-making in the world. It is also about how existing rules of the game governing what counts as meaningful, as sensible, as verifiable, are being upended. The human subjects of the data-driven society do not live blinded and walled off from the world of machines; rather, they are constantly asked to interpret its emissions, acquiesce to its findings, respond to its insights, and cope with its judgments. The gap between the human subject and the world out there, between the thinking, feeling self and the body in here, is further divided and exploited through this advent of data as non-sense.

In seeking to understand the basic architecture of human embodied experience, Merleau-Ponty spoke of two kinds of senses: *sens*—meaning "sense," "direction"—and *le sentir*—"to sense," "to feel," sometimes translated as "sense experience."[7] Today, everyday English retains a double meaning for *sense*: a sense for feeling and a sense for making something meaningful. Similarly, these stories point toward two kinds of shifts in what we sense and what sense is made of it in the data-driven society. The first concerns the differences between what human and machinic sensibility can perceive and process; the second involves how that information is then placed into different contexts for meaning-making, such as the relentlessly indifferent correlations of big data analytics. Not only is this new knowledge alien to human sensibility and cognition, but it also often begins operating toward "efficient" outcomes that have little regard for human experience. If big data began with a technical kind of indifference toward causality or theory, its applications increasingly exhibit an indifference toward human priorities as well. Ian Bogost warns that as smarter machines fill the spaces in and around human lives, "technology's and humanity's goals [will] split from one another."[8]

Data-sense, as a vision of technological augmentation and posthuman upgrade, is thus founded in observations about the non-sensicality of data and the incomprehensibility of machines. Even as the increasingly systematic extraction of personal data for commercial uses provokes concerns that "we are [becoming] strangers to our normatively aggregated selves," the same alienness of machinic sensibility inspires commercially attractive proposals for the measurable human.[9] To be clear, data-sense rarely appears as such a consolidated and general theory in the public discourse around datafication and self-surveillance.

This is an analytical reconstruction of the recurrent patterns in how the benefits and effects of these technologies are framed, how they mobilize human subjects into what trajectory of transformation, and how they position present technologies in a broader historical narrative of progress and objectivity. We come full circle back to Weizenbaum's dictum on the reshaping of humans in the image of technology. If machinic sensibility describes smart sensors' much-vaunted ability to collect data that humans cannot, data-sense describes the ways in which humans are to be asked to become interfaces and clearinghouses for that data.[10]

Posthuman Subjects

Although data-driven fabrications produce knowledge that is unconcerned with human meaning-making, this does not imply that human subjects are simply left alone. On the contrary, the growing importance of smart machines demands specific virtues, skills, and attitudes for the tracking/tracked subjects. At the most basic level, we find the idea of data-sense as a certain literacy. In 2012, *Wired*—ever the evangelist for new practices in computing and whose employees had co-founded the Quantified Self (QS) community—hosted a conference subtitled "Living by Numbers." On the podium was Kevin Kelly, the co-founder of QS, and he did not hesitate to sketch the big picture for tracking technologies:[11]

> We're horribly, I mean, we're just not evolved to deal with numbers. Our brains aren't really good with dealing with numbers, we don't do statistics very well, we're not really a number animal.

Here, the narrative remains at a utilitarian and instrumental level: self-surveillance demands new skills that humans can learn (with technological support), akin to the problem of learning the right grammar or typesetting. QSers and the self-surveillance industry have long grappled with data visualization techniques as a way to render numbers into patterns, juxtapositions, and curves—objects that communicate a certain narrative form. (An online tool developed by Intel Labs, itself named Data Sense, offers a secondary communicative layer to the proliferation of machine-gathered data, allowing users to import, aggregate, and visualize data they have gathered through tracking devices.) As a broader vision, however, data-sense often went further. Having begun in terms of literacy or numeracy, Kelly went on to argue that machinic sensibility can and should be appropriated into subjects' own sensory experience. In such a future, users would no longer have to struggle with the "raw" data or even the proliferation of graphs and tables but become so used to having machinic communications all around them that they would intuitively and habitually reach for input from tracking devices to check whether they are hungry, where they are going, and whether they are properly rested.[12]

> But what I think the long term direction of this is, is, we want to use these sensors we're talking about to give us new senses. To equip us with new ways to hear our body . . . right now we have to see the data, the charts, the curves, but in the long term where we want to go is, we want to be able to feel, or see, or hear them.

These "new senses" are thus described as an internalization of machinic temporalities, rhythms, and patterns of communication into the tracking subject's phenomenological equipment. Chris Dancy, the "world's most connected man," testifies that his experiences with tracking had indeed left him with a kind of data-sense: "I no longer need sensors, I realise I am a sensor."[13] Dancy claimed that over time, he had internalized the machinic sensibility of his devices to the extent that he could sense the temperature, and other forms of data around him, with an unnaturally high accuracy—and that he is even able to diagnose other humans on the spot, using a few quick observations to reach accurate estimates about their sleep and other activities. Such testimonies were not unusual within and around the QS community.[14] One entrepreneur, for whom self-surveillance became a way to develop a more precise awareness of her health problems (as well as her data-driven health coaching business), described to me how that experience had equipped her to diagnose the hidden correlations that trouble other bodies as well:[15]

> I said to [my friend,] you are aware that you are dairy intolerant, right? And he was, like, what do you mean, I'm dairy intolerant? And I'm like, yeah, you have a little bit of a . . . sound on your throat when you eat

dairy . . . I'm pretty sure you're dairy intolerant. And he says, that is the weirdest thing, I'm Indian!

It's interesting that you're able to read somebody else now, even though, technically, you don't have any access to his raw data.

Oh, totally. Totally. It takes me fifteen minutes on airplanes to tell people what it is [that ails them].

Once again, health and medicine represent with the greatest clarity this tendency for magnifying, fragmenting, and expanding the body into ever-larger databases and points of measurement. One point of overlap with self-surveillance involved the popular fascination around the microbiome: the ecosystem of bacteria, fungi, and other organisms that reside in the human body, especially the gut. Typically credited to molecular biologist and Nobel Prize winner Joshua Lederberg, the microbiome has become a popular object of fascination and hope about breakthroughs in human health, both inside and outside the scientific community. In the world of self-surveillance, the microbiome emerged as a particularly potent depiction of the necessity of machinic sensibility. Here was an entire system latent in every human yet entirely inaccessible to human sensibility. Its composition is unique to every individual, favoring the "n = 1" approach of self-trackers over populational norms or the Queteletian *l'homme moyen*. Self-surveillance discourse thus latched onto the interest surrounding the microbiome, describing it as an untapped lode of new knowledge just waiting to be colonized by tracking tools.

This narrative received an appropriately iconic protagonist in Larry Smarr, a physicist who masterminded the foundation of America's Supercomputer Centers Program in the 1980s. Smarr's personal efforts to track his health throughout the 2000s, it was claimed, allowed him to correctly diagnose the onset of Crohn's disease.[16] By the early 2010s, he had constructed a chamber at Calit2, his home institution. There, the walls were lined with screens providing a walk-in 360-degree view of his gut microbiota.[17] Meanwhile, Smarr's experiments had led him to the QS community, which saw in him a pioneering and ambitious example of where self-surveillance could lead us to. Although Smarr's claim to concrete, specific achievement in health knowledge required an extraordinary range of personal expertise, institutional backing, and other unusual resources, enthusiasts could point to his case as a way to bridge the relatively mundane and often failure-prone devices of self-surveillance's present with the rhetoric of its transformative future.

Such imagined reconfiguration of the subjects' sensory equipment reached their rhetorical peak in claims of a broader posthuman shift.[18] Kevin Kelly terms it *exoself*: an "extended connected self" that constantly discharges data while receiving all kinds of machinic communications, both consciously and nonconsciously.[19] Such language extends the long narrative of technological transcendence that had characterized utopian (and some dystopian) rhetoric about personal computing and the internet in previous decades—and, indeed, the broad and powerful influence of cybernetics throughout the twentieth century that defined the body and the selves as information systems.[20] One particularly relevant branch of that cybernetic imaginary had been the countercultural influence on personal computing and the Internet as a route to a technologically expanded consciousness—a vision that Kelly himself had actively brokered in the 1980s and 1990s.[21] In this tradition, tracking enthusiasts across news media, internet technology industries, and the QS community spoke of exoselves and new senses. Such discourse promised not a future where users are turned into hyperrational machines but, rather, a more "authentic" relationship to one's humanity.

There are, of course, vast chasms between the promised future of data-sense and the present technological practices supposed to represent society's progress toward them. Spectacular cases such as Smarr's tend to be singular and rare. To be sure, opining that present technologies do not exactly fulfill the nebula of promises surrounding them is somewhat beside the point. As an amalgam of fantastic rhetoric, material glimpses, and anticipatory actions, data-sense is fundamentally about overstepping the reality of existing achievements and orienting public imagination toward a particular future. What matters, one might say, is the ability to write the future, *a* future, on the basis of whatever the present has to offer—enough that the rest of us might feel sufficiently enthused, compelled, and anxious to get on board. The many discursive and material presentations of data-sense thus coalesce around a simple and universal ambition: to reach beyond the horizons of human phenomenology and to become more compatible with our machines who have already journeyed far beyond the human senses.

Modding the Senses

In all this, data-sense reflects the broader societal interest in the nonconscious. Although one cannot consciously "think" one's own nonconscious, this layer nevertheless organizes information from the external world for conscious operations. By "synthesising sensory inputs so they appear consistent across time and space, processing information much faster than can consciousness, recognising patterns too complex and subtle for consciousness to discern,"[22] the nonconscious reveals the mediation inherent in every form of conscious "experience"—a term that itself reflects an entire mythology about the integrity and non-mediated naturalness of human sensibility.[23] As Katherine Hayles has shown, the growing focus on the nonconscious over the later twentieth and early twenty-first centuries has owed much to the popularization of neuroscience, as well as proximate and sometimes precursor developments in areas such as cognitive biology and cybernetics.[24] If phenomenologists a century earlier had riffed off gestalt psychology and other contemporary sciences to theorize how humans make sense of the world around them, the turn toward the nonconscious targets a broader problem of how human, nonhuman, and even nonsentient technical objects operate through "nonconscious cognition" (NC): the contextual interpretation of information that does not require conscious thought or self-awareness.[25]

Technologies of self-surveillance are, in many ways, technologies of and for the nonconscious. In tracking the body's stress levels to manage concentration and productivity, or the idea that human users will eventually develop expanded capacities to sense the world, we find this recognition that the nonconscious features a set of powerful parameters governing the limits of conscious activity—and an ambition to codify and modify those parameters through technology. After all, if human consciousness is managed by a set of nonconscious systems that protects it from becoming overwhelmed and helps it sort the world into sensible phenomena, then it is exactly that layer that tracking devices promise to calculate and optimize. Data-sense can be understood here as a meeting point between human biological NC and technical NC at the contested terrain of the human body. As a quasi-evolutionary fantasy, data-sense describes a technological provocation of the human nonconscious towards a more hybrid and expanded state. If existentialist phenomenology had sought the basic and often pre-conscious mechanisms of human being-in-the-world, that pursuit returns full force in the data-driven society. True to the pragmatic philosophy of big data, it is focused on the optimal modification of these mechanisms rather than any discovery of their fundamental roots.

Such modification requires not only the deployment of ever-smarter machines, ever savvier to the idiosyncrasies of their human subjects, but also the latter's ability to *adapt toward* machinic sensibility as well. The body as the object of surveillance and datafication (the body that speaks), the body as the material terrain of the nonconscious, undergoes constant reprogramming somewhere between conscious reason and the mute biology of the flesh. In tune with the broader popularity of "hacking" the body, tracking devices are increasingly cultivating fresh channels for communicating machinic data to human subjects.[26] These channels are often designed not so much for discrete queries and deliberations but a habituated and tacit receptivity to the continuous flow of machinic communications. To sense one's own body is already for consciousness to direct itself toward something else, an object; self-knowledge, in other words, is to know the "me" as an object distinct from the conscious "I."[27] Self-surveillance is designed precisely to intercede in this relation.

Here, what matters beyond the design parameters of any single tracking device is the broader expectation that when we seek knowledge of the body, we shall turn to smart machines and learn to see ourselves in terms of databases and correlative predictions.[28] As I use a tracking device to train better habits for sleeping or eating, I also become habituated into the relationship of machinic supervision, of seeing myself through the mirror of the sleep score. In the absence of more direct relations of coercion, what is at stake is not necessarily the standard metric presented by a tracking device but the gradual development of the subject's *awareness* of their own body in datafied terms. Data-sense describes these decentralized efforts to write our collective future, a future in which the conditions governing how we "experience" our bodies, make sense of our own emotions, and produce knowledge that counts for who we think we are are gradually aligned to conform to the capabilities of machinic sensibilities.

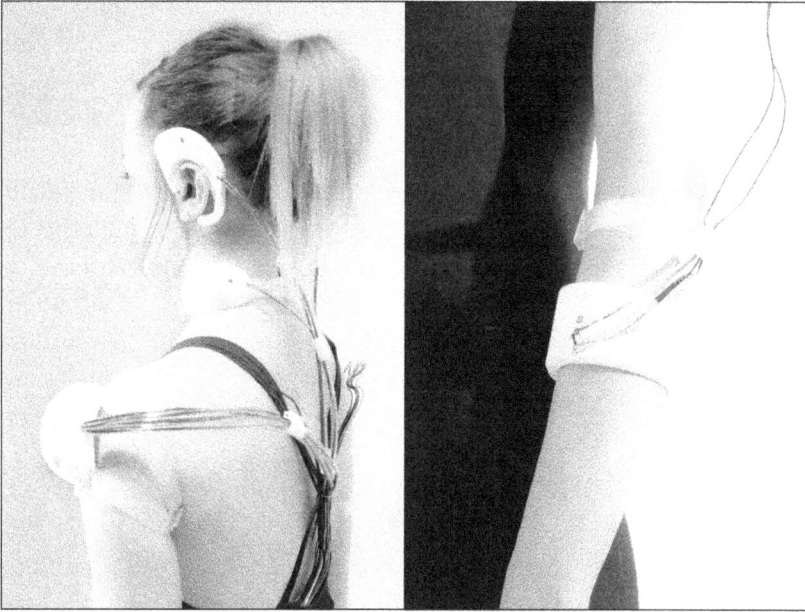

Figure 6.2. Ling Tan explains that the experimental device "investigates the limits of human bearability towards wearable device[s]."
Source: Ling Tan, "Reality Mediators," November 22, 2015, http://lingql.com/reality-mediators/.

One sensory frontier for these adaptations is the haptic/kinesthetic. Touch is a sense not so exhaustively overcoded with theory as vision has been in the West and is a channel that is difficult for the body to suppress or ignore. As such, tracking technologies that exploit touch have focused on its promise of affecting individuals in ways that cannot easily be denied or resisted. In 2013, designer Ling Tan's Reality Mediators were nominated for the 2013/14 Internet of Things awards and were duly covered across media outlets (figure 6.2). The device—a bare set of cords, fasteners, and white modules—is designed to apply discomforting and painful feedback, including the ability to "zap" the wearer if the device detects signs of flagging attention levels. Around the same time, two MIT students devised Pavlov Poke, a simple combination of existing computer activity trackers and basic electric circuits to send shocks to the user if they should be distracted from productive work. Of course, the specific form of distraction that inspired Pavlov Poke was itself a case of the technologically reprogrammed nonconscious; the creators explain that Facebook browsing had become so habituated that they would often be "dragged there by some mysterious Ouija-esque compulsion."[29] One type of nonconscious adaptation called for another.

Such forms of haptic discipline—enacted not through language, image, or conscious negotiation but through targeting the involuntary openness of the senses—were also making their way into commercially available products. Productivity hacker Maneesh Sethi claims that he invented the Pavlok after hiring a human agent using Craigslist to slap Sethi in the face whenever he was distracted.[30] Because the human component was not easily scalable, it had to be replaced; the resulting product consisted of a smart wristband that delivers electric shocks to inculcate desirable habits—such as zapping users to help them wake up on time.

Such reprogramming of the self unsettles the political and moral stakes of what we experience and how we experience it. Smart machines entail not only the expanded manipulability of the conditions of human experience but also *externalize* this manipulation unto the realm of the social. Human sensibility depends more and more not only on its unconscious architecture embedded into the body but also on an array of external objects. The latter's multifaceted social life as technologies, as black boxes, as commercial products, and as anthropomorphized agents, exposes them to conscious human negotiation over how exactly their preconsciousness might be engineered.[31] Wittgenstein's groundless ground perseveres but is increasingly explicitly recognized as a technologically augmented simulation rather than any biologically or philosophically essential foundation. Externalization opens up the nonconscious to divergent reprogramming along lines such as socioeconomic status or normative ideas of mental illness. Different kinds of bodies are, once again, sorted into different forms of factmaking. In the world of data-sense, who—or rather, what combination of technological objects, human design, and systemic tendencies—will determine the truth of a body? Which machines, and behind them, whose technological designs, which kinds of numerical tendencies, will count as my truth—no matter what I might have to say about it?

Future Perfect

Even as data-sense presented a universalizing vision for hacking human sensibility, only a few limited functions were actually available—and typically only for the most tech-savvy sliver of the wider population. From James Proud's "glorified alarm clock" in chapter 1 to Kevin Kelly's direction-seeking belt, we find a consistent gap between the brilliant future and its pale shadow of a present. These fissures are sutured by a kind of futurism more than familiar to the children of the Enlightenment. Insofar as data-sense is presented as an inevitability, the tracking/tracked subject is urged to develop a new set of skills, ethics, and attitudes to get the most out of new technologies. Such a narrative enlisted a highly celebratory history of computing technologies in the twentieth century. Self-surveillance was often presented to the public as the next step on the march of technological progress, with dedicated QSers its vanguard:[32]

> One of the mantras around Quantified Self is that obsessive self-trackers may look outrageously geeky now, but they will soon be the new normal . . . we all will be living in an ocean of data in the near future, whether we are self-tracking or not, and learning how to read, manage, retrieve, understand, digest, parse, and selectively ignore this flood of data will be an essential skill—for individuals and for organizations. Self-trackers are there first.

Such discourse anticipates that new, or even as yet uninvented, technologies might furnish a sensory and experiential *default*.[33] This future-painting coincides with the "new normal," a term observers of September 11 used to "signify a world destabilised by terrorism, economic fluctuations, and contagion prevention."[34] The new normal is a very much presentist description of what contemporaries felt to be a radical and even unprecedented disruption. Here, technology is unmoored from the many possible human choices that regulate the path of its development and elevated to an impersonal tidal force directing civilizational progress—exactly the sentiment that inspired the modern meaning of the word *technology* in the first place.[35] In the aftermath of September 11, state surveillance and intelligence actors claimed that a newly dangerous and unpredictable geopolitical reality demanded the development of powerful and indiscriminate electronic surveillance. And as we saw in chapter 4, self-surveillance technology was often embraced as an "obligatory technology."[36] It might as well have been a Kafka joke: it has been decided that society is going to become data-driven, and we are all going to become self-trackers, and although nobody's quite sure who decided anything when, we had best not be left behind.

To keep up with this future, the individual is asked to keep upgrading. In keeping with the computational metaphors, the human subject is seen to provide an underlying set of functions and frameworks; memory, vision, cognition, reason . . . and these capabilities, it is argued, can be newly programmed and augmented to stay compatible with increasingly sophisticated technological prostheses. Data-sense functions as a "technological habitus": the social tempering of the body that allows individuals to internalize the virtues, senses, and know-how necessary to prosper in that technological society.[37] Through this process, subjects are promised to achieve the kind of efficacy or competency that "counts" in the new regime of knowledge, whether it be the savvy social media user that extracts reputational and monetary rewards through the network or the self-surveillance connoisseur who can comprehend and manipulate the new wealth of personal data. Just as state surveillance seeks unprecedented transparency on the part of the population to read its criminal intent, self-surveillance's ability to make one's own body more transparent to oneself produces new grids of legibility for employers or advertisers. The project of knowing more about and optimizing the self enrolls many different commercial business and government interests.

Data-sense thus describes the widespread discourse that exhorted the necessity of a new kind of user-subject. Analogous to the creation of the computational user[38] or the netizen, this interpellation involves a sustained effort at codifying and organizing the masses to adopt new epistemic processes, new ways of seeing, that would render them more amenable to the strategies of data-driven monetization and governance we have witnessed in chapter 4. Again, the future sketched out in data-sense is a future far too sweeping, far too consistent, far too optimistic to expect perfect fulfillment. Yet these cultural fantasies are playing a game of prophecy, not prediction. Their significance lies in their ability to help grease the wheels of technological deployment in their present—stitching together exercise-tracking wristbands and direction-sensing belts, mood trackers, and sex quantification apps into a broader fabric of meaning and belief.

The messy knot of prototyped ambition, ubiquitous product, and futures sketched boldly on PowerPoint slides: data-sense illustrates a powerful future-forwarding tendency in our technological fantasies, a tendency that draws its affective force from past generations of the pursuit of objectivity. In chapter 1, I characterized this tendency as a honeymoon objectivity: the recurring faith that this time, the new generation of technologies will fulfill the epistemic fantasy—a pattern that helps self-surveillance borrow from the cultural legitimacy of older machines and institutions. As strategic and canny as this historicization is, it should not be mistaken for a truly consistent and universal project of a singular objectivity. Self-surveillance involves many different technological practices, yielding no clear definition of what is objective and what is "good" knowledge. This ambiguity benefits, rather than hinders, the technological fantasy. After all, tracking involves a wide variety of actors and interests at stake, from the QS to mass-market product developers, mainstream media commentary, and the consuming public. In this disaggregated landscape, objectivity often retains its pulling power by appearing not so much as a perfect and singular ideal to strive toward but as a relatively generous category that describes a set of general dispositions toward knowledge and its uses. Gary Wolf thus suggests that what matters for the community is not any codified dogma of objectivity but certain styles of putting facts together:[39]

One is objective means explicit. And you could say formal, in that sense. Numerical. So if you think of it in terms of representation, you could classify types of representation from formal to informal . . . Another is social versus individual, in which individual perception would be considered subjective, and social—an idea, a perception, an observation that entailed agreement of more people, whether that be expert observers or everybody, would be objective . . . I think that [the two definitions] probably have to come apart to understand the Quantified Self, and the objectivity of the Quantified Self really refers to the way it's actually used. It refers to [the] formality of expression, and not so much to a kind of social validity.

Again, we find a certain pragmatism toward truth and knowledge that characterizes the wider turn to big data. A measure such as self-reported mood scores may never attain a perspectival objectivity, and its actual fidelity with the raw empirical phenomenon in question might remain uncertain as well. What matters is to produce comparable, calculable, and transportable measurements—things that you can operationalize and instrumentalize.

Such folk theories hardly add up to a comprehensive philosophy, but they do not need to. Their contribution lies in helping develop robust connections between different interested parties and discourses, enrolling a wider variety of subjects into self-surveillance and the promise of data-sense. One part of this effort has been the gradual weaning of the self-nominated "Quantified" prefix: a discursive disaggregation of the virtues of formal, standardized, machinic objectivity from the stereotypes of inhumanly hyperrational number geeks. When Wolf wrote for *The New York Times* in 2010 to introduce the nascent QS culture to the wider public, one of his strategies was to cite Charles Dickens's Mr. Gradgrind.[40] Where Gradgrind was an insufferably "obstinate" stiff who insisted his pupils chant, "Fact, fact, fact!" Wolf insisted that "it is normal to seek data." In this vision, self-surveillance pursues objectivity not as a dry and scholarly insistence on precise factuality but as a practical virtue that seeks the eradication of biases, errors, and other forms of epistemic uncertainty when it comes to self-knowledge.

This transposition of the virtues and legitimacy of objectivity also extends to ideas about the normative human subject of data-sense. If objectivity in many contexts developed a suspicion of subjective claims and often sought to inoculate its procedures from human tampering, this also meant strictly prescribing the kinds of subjective virtues and practices required to ensure such separation. Daston and Galison describe how mechanical objectivity involved its own kind of self-surveillance: scientists were required to diligently curb their human temptation to subjective bias and projection. The ideal subject of this epistemic regime would record exactly what the camera or microscope sees—a patient and diligent work of suppressing the subjective that Daston and Galison compare to Schopenhauer's "will to willessness."[41] The subjective virtues of objective epistemology was not limited to the art of disappearance, however. Scientists were also expected to exhibit a set of moral traits, which ironically included the ability to confidently *imagine* a utopian future for their particular style of fabrications.[42] At a practical level, where the work of machines is inevitably assisted by human hands and eyes, the ideals of objectivity translated—at least in the increasingly institutionalized and professionalized halls of "normal" science—into the valorization of qualities such as reliable and dependable personalities.[43] To articulate the virtues of objective knowledge is to specify a subjective ethic that cultivates sufficiently disciplined practitioners. In this context, the rhetoric of data-sense leveraged self-surveillance's embrace of objectivity toward its ideals of data literacy, exoselves, and the "new normal."

Data's objectivity, then, operates as a regulatory ideal: a distant focal point that organizes a collective orientation. The future best exerts its gravitational force when deferred onto the horizon. Self-surveillance's selective appropriation of objectivity reflects the long historical collaboration between the modern ideas of technology, progress, and objectivity. If objectivity is the regulatory ideal bestowing a certain picture of the world and the moral virtues associated with the pursuit of knowledge, technology is the name for an infinite array of solutions that are supposed to constantly create new domains of action to replenish this pursuit. As we saw in chapter 1, the rise of mechanical objectivity in the nineteenth century was provoked by the development of new photographic technologies. Indeed, technology as an idea brought to the table a host of narratives about the virtues of such epistemic projects. The modern meaning of the term itself emerged in the mid-nineteenth century[44]—a moment when earlier designations of the practical arts and crafts were disrupted by breakthroughs in industrial machinery. Just as contemporaries of datafication speak of the coming ubiquity of new technologies, early commentators of technology perceived a proliferation of machines that reflected an unprecedented rate of invention and progress. It was an understandable sentiment, given the extraordinary expansion of material wealth and mechanical infrastructure Europe and America achieved during the Industrial Revolution.[45] If the "mechanical arts" referred to specific practices of tinkering and craftsmanship, the accession of *technology* as a term marked a new belief in a broad, hyperobjective system, such as the railroad.[46] Theories of the autonomy of technology, and its overlapping with society itself, would follow later.[47] Finally, this sense of a legible and countable world was able to extend and renew itself through the idea of progress. Although there is a long buildup of various partial theories about human betterment from the Renaissance onward, again only during the eighteenth and nineteenth centuries did the now-familiar package come together: a belief in the inevitable and indefinite betterment of civilization through technical inventions.[48]

Self-surveillance's futuristic vision of automated self-knowledge rides the coattails of this long historical wave. Here, we must understand the function of "technology" as a concept, as an image, as distinct from the deployment of actual and specific technologies.[49] One consequence of the increasing complexity in technological systems was that "the public increasingly had to rely on *images*"—that is, narratives and emblems for popular consumption and imagination—"of technology for its understanding of both technology and progress."[50] At this level, data-sense profits from the image of technology as a neutral solution that can be applied to every problem.[51] Once again, corporate data mining forged a path that self-surveillance would follow. Between mid-2000s and the mid-2010s, social media platforms built up a powerful position of authority as institutions for the circulation of both human sociability and commercial sales of personal data based on that sociability. They did so precisely through this image of neutrality, which ultimately helped reroute massive social and commercial influence unto a select few media corporations.[52] Here, claims of empowerment and popular sovereignty became the unwitting dummy for the surveillance and exploitation of personal data.[53] Silicon Valley, the epicenter of this strategic fiction, framed its role as architects of *disruption*. The word expresses a certain indifference to existing social contracts, to existing margins of tacit knowledge and unofficial relationships, in favor of universalizing technical principles: accuracy, efficiency, and optimization. This rather violent conception of technology again draws its legitimacy from the apparent universality of objectivity and the inevitability of progress—ensuring future honeymoons to come.

The imagined futures of data-sense were christened through their own collection of images and metaphors. By the 1990s, the popularization of the internet occasioned metaphors of cyberspace, virtual communities,[54] and the "information superhighway."[55] Self-surveillance itself tapped into a set of metaphors governing the broader fascination with (big) data and its wireless circulation. This included the "cloud"—a vision of smooth, immaterial technology, free of political and physical obstacles and the vision of a certain epistemic proliferation, all of which forms a "topography or architecture of our own desire."[56] Personal data itself received new and often grandiose descriptions: "data is the new oil" was the buzz phrase in the mid-2010s, expressing again the universalizing sentiment that every phenomenon in our society could be colonized through technologies of datafication.[57] In this language, data is a ubiquitous, amorphous good whose discovery and extraction can be engineered through new technologies—never mind the ecological devastation that this pursuit of fossil fuels has now wrought on the planet. And for self-surveillance, too, this datafication is at the crux of the promise of technological objectivity.[58]

In short, self-surveillance leverages historically accumulated virtues of objectivity and technology toward a vision of neutral optimization and data-driven truth. It not only repeats earlier waves of hyperbole surrounding new technologies from electricity to the internet but also deliberately reproduces such fantasies as part of its bid for social significance and public uptake. Again, such promises of technological transformation are rarely fulfilled in exactly the ways foreseen by its pioneers. Self-surveillance is unlikely to eliminate uncertainty in self-knowledge any more than mechanical objectivity could eradicate the scientist from science. We may look forward to a time when today's smart machines, filled with such hopes and fears, itself becomes the stepping-stone for the next sociotechnical revelation; as if this time, the Hegelian dialectic will come to a conclusion, and pure objectivity might finally be at hand. Honeymoon objectivity is no prophecy but works very much in the present, legitimizing new power relations and systems of veridical authority demanded by new technological practices.

Control Creep

If data-sense is a case of honeymoon objectivity, if it reprises and modifies the Enlightenment orientation towards objective truth as a grounding for knowledge and social order, if it is the latest chapter in technologies as the containers of our desires—then what kinds of politics is to be had in the meantime? However ephemeral the images of the future may be, they ultimately come to bear on real bodies and lived experience. The question remains: What kinds of control, nudging, and cultivation is effected on human subjects toward what forms of social organization, entailing what distribution of rights and responsibilities, knowns and unknowns, the visible and the invisible?

During the early twenty-first century, big data and smart machines achieved the market penetration and notoriety that come hand in hand with their entry into the normative fabric of society. The Snowden affair, of course, was one such landmark. In its wake, the public was asked to undertake a collective intellectual exercise: to shift their styles of reasoning from a model of direct causal relationships between discrete, intentional actors—wherein a centralized and totalitarian "Big Brother" watches the people, for instance—to a more complex network of subjunctive situations and preempted dangers. The concerned citizen was asked to think not of the particular harm done the individual through the malice or incompetence of specific government agents but of a speculatively couched combination of possible and contingent risks. The threat to national security, too, shifted from the imagination of clearly defined enemies to the amorphous, statistical danger of "the whole haystack."

This messy backstage of datafication has provoked theories of a softer, subtler, more roundabout kind of biopolitics: techniques of control that veer away from direct contact with the individual and their lived experience but operate from a mediated distance, manipulating the categories and standards by which people come to count.[59] Such analysis extends what Foucault considered the basic principles of control in a liberal society: political subjects behave according to a certain reason, and government is the art of discovering and manipulating those parameters, leading to forms of control-at-a-distance.[60] But if the data-driven society is a society of softer biopolitics, what does that mean for existing conceptions of freedom, of choice, of resistance? We saw how the valorization of transparency entails a new distribution of labor and responsibility, a greater pressure for the individual subject to stay informed, to stay up-to-date. We also saw how these new burdens pass under the radar when the shifts are conceptualized in more traditionally liberal language. If the means of control have changed, then so should the definition of resistance or empowerment.

In analyses of power and control in the new media society, the five pages of Gilles Deleuze has served as the master manual: the "Postscript on the Societies of Control."[61] The central lesson of that brief note was that control is no longer achieved through discrete objects bestowed with certain specific powers but modular and substitutable objects enacting a certain relationality. One is asked to visualize not a relationship between one being and another, each on its own ground fighting for territory, but the ways in which the individual is prefigured, prepared for, primed, preconstituted, predisposed toward certain patterns of desire. This contrast has been embraced as an expression of what is truly novel about late capitalism and new media—although any idea of a clean historical passage from Foucauldian "discipline" to "control" relies on a poor caricature of those concepts.[62] Away from the distracting question of historical novelty, what matters is how a new generation of subjects are being prepared for their particular, technologically dominated form of life: a way of living that includes how we learn to conduct our bodies, to perceive them, to trust and distrust our consciousness, to think of our thoughts and knowledge and ideas as joining what kind of larger collective flows and how the limits of our knowledge and the limits of our actionable world become delineated.

It goes without saying that such cultivation does not emerge from any central planning committee. Big data and smart machines are not inevitable extensions of neoliberalism,[63] nor is the technology inherently the harbinger of unhappy, alienated subjects constantly torn between the expectations of quantified optimization and the uncanniness of their own experience. But their historical development has resulted in technologies that target the nonconscious gap between human experience and the world—technologies that construct new patterns and boundaries of intelligibility. In other words, surveillance and tracking systems are establishing a "different set of power and control nodes,"[64] including on processes that used to remain relatively unofficial and/or self-driven.[65] In response, the political and ethical stakes must turn to what kinds of rules of the game these technologies are being made to support and what spaces for alternatives remain.

In particular, the forms of control exhibited by technologies of surveillance and tracking throw into disarray the long-standing liberal emphasis on individual, informed choice as the foundation of morality and politics. From the outsourced labors of a transparent society to the increased "maintenance work" demanded of tracking subjects, the equation of knowledge with agency and empowerment is superseded by new bargains in which knowing is often the beginning of new labors. The exercise of choice, so central to the empowering promise of self-surveillance (and, to a lesser extent, the debates around data privacy and transparency), also sits uneasily on the axis of control and freedom. Even as these technologies are tasked with datafying the everyday for the willing consumer/user, the leakage of these functions out into workplaces, institutions, the "Quantified Us," speaks to their participation in the growth of the data market. The private, personal, domestic dimensions of the body, so much of which often remained "known" in unofficial and unsystematic ways, are being colonized by a datafication process that renders them compatible with the calculations and judgments of the market and of the government. The very mundanity of these products point to a certain level of normalization; datafication as a logic of governance reaches beyond the dramatic moments of individual lives and into the many little choices that make up the most ordinary of days.[66] As smart alarms turn on our conscious cognition, as smart forks advise in real time that the next mouthful exceeds the calorie quota, as smartwatches congratulate us for standing up every hour, the everyday labor of choosing, consuming, and knowing becomes less a space of nondeterministic creativity and more the work of further aligning oneself to the regime of what counts as truth.

In January 2017, *Wired* published a special issue dedicated to science fiction, explaining that "sometimes to get a clearer sense of reality, you have to take some time to dream."[67] One of the featured stories depicted a world saturated with the choices, choices stripped of politics: the work of rating and polling becomes a ceaseless, inane rhythm of life in a world entirely governed by A/B testing.[68] "Which word feels sadder: *lonely* or *lonesome?*" What about a local bike-lane ordinance, mustard offerings in fast-food restaurants, or the ideal temperature for hot chocolate? These many little choices become the surest justification for the world around us: because every single thing is the product of an instant poll, an online petition, a Twitter brainstorm, a Reddit upvote, it *must surely* have democratic legitimacy. At the same time, the labor of such "choice" produces a Borgesian virtuality. In the story, the full-time chooser spends all their day voting, with never a moment to glance at the world unmediated. This corruption of liberal choice extends to the domain of facts: "65 percent of the public wants the government to do something about the Great Midwestern Drought, though 68 percent of the public also believe the drought is a hoax." How can the subject forge alternative forms of living or cultivate one's own meaning and place in the world, when we are kept so busy with all this technology and when the work of optimizing ourselves takes up so many of our days?

The Body at Stake

The many little choices of the data-driven society do not mesh well with any clean separation of power and resistance, structure and agency. The portraits of its human subjects are no longer filled with the prisoner in the cell, the proletariat at the factory lines, the diligent fellow burning the midnight oil with a Protestant ethic. We find instead the knowledge worker, coping with irregular, project-based conditions of employment, coping with the new labors necessary to retain the kind of marketability that counts in the new economy.[69] Alongside them is the concerned citizen, devouring (or devoured by) the inexhaustible stream of leaks and revelations, paralyzed by paranoia—a paralysis that sucks up so much time, so much emotion, to maintain. The critical object of politics here is not *knowledge*, fantasized as a relatively immaterial problem of educated minds, indefinite archives, and near-instant transmission. Rather, it is the panoply of fabrications, the mundane everyday heuristics that cobble together a sense of a coherent society, manageable limits of the knowable, and workable divisions between humans and machines. What must I know about myself, and render legible to others, to remain eligible as a worker, a dating prospect, an ordinary individual? How must I figure out ways to ignore, skim, defer, pretend, assume, and believe in response to all the knowing that is expected of me?

Lone wolves, quantified selves: these are bodies that speak, bodies that are *made* to speak as a function of power. Amid the rhetoric of cyborgs and smart machines, virtual reality and simulations, the human body and human experience remain the contested terrain for questions of power and politics. The body has been called the "first instrument": human sensibility, despite all its modification through technology, remains the always-already layer of mediation, one which technology surpasses but cannot leave behind.[70] This is not to posit some essential and universal sensibility untouchable by social or technological forces but the exact opposite; it is precisely because we cannot get away from our bodies that technology feverishly intrudes on it. Theories of nonconscious cognition and the bypassing of the human sensorium have their pulse exactly on this effort to construct a new communicative circuit for truth production, one that hopes to optimize the traffic between body and machine (and thus to the cloud, the institution, the database, the document). We continue to dream of bodies that might be read like open books.

But who, or what, is taking on this task of rendering bodies into words, numbers, and data? So much of this story has been not only about technology but also about everything about technology that goes beyond the machine, the code, and the interface. Datafication produces no dimension of epistemic purity from which depoliticized facts may be imported to clear up the messiness of the social and the political. From the justifications for preemptive arrests of terror suspects to the pursuit of new secrets for optimizing individuals' productivity (by themselves and by their employers), the very invocation of data entails a human–nonhuman network of actualized practices, aspirational horizons, posthuman rhetoric, and particular ways of talking and thinking about bodies and knowledge. In short, data-driven objectivity describes not merely a set of technological functions but also an *institution*. Not an institution in the sense of a centralized and intentional actor that might wage war on human freedoms. There is no human subject who stands apart to wage this war, and neither is there any truly autonomous machine on the other side of the trenches. Rather, datafication as an institution points to the emerging norms and rules by which the calculation and circulation of datafied bodies are brokered. As technologies of datafication aspire to a new normal, they are positioned to disrupt not only the commercial structures of communication but also the phenomenological default of how we use our bodies and how we experience them. Self-surveillance represents a particularly mundane and forgettable level at which datafication infiltrates the conditions of human experience, allowing new forms of external manipulation by humans and nonhumans alike. It has to do, as we said at the very start, with the molding of *desire* itself as a social structure.

Conclusion

What Counts?

[T]here are certain tasks which computers *ought* not be
made to do, independent of whether computers *can* be made
to do them.
—Joseph Weizenbaum, *Computer Power and Human Reason*

What counts as knowledge in a data-driven society?[1] What *should*
count as legitimate ways of converting bodies into facts? The pursuit of
machinic objectivity calls on human subjects to know more than they
reasonably can and, in their failure to keep up, defer to new institutional
arrangements of recessive and opaque technologies. The posthuman
fantasies of data-sense increasingly normalize an interminable labor of
datafying the self, through which my sovereignty over the truth of who
I am is externalized onto smart machines and corporations. The question is made more difficult by the undeniable benefits of technologies
of datafication. Big data certainly boasts impressive predictive power, at
least under the right conditions. Smart machines offer an alien sensibility, one that is able to produce the kind of knowledge the human senses
are not equipped for. But can technological reason provide normative
guidance on how we should engage (and disengage) with it?

The trouble is that, even as big data aspires to the epistemic purity of
machinic objectivity, bodies constantly exceed data. The fabrications of
datafication provide not irrefutable certainty but a new collection of mediated, deferred, imagined, assumed ways of feeling like we know, like
we are in control, like we can do something about the way of the world.
Forms of uncertainty—as striated horizons of recessivity, as subjunctive conditions, as interpassive beliefs—hold together the "better knowledge" of the data-driven society, even as bodies continue to exceed and
confound datafication.

The trouble is not that datafication is imperfect but that it opens up a gap between the futuristic promise of better knowledge and the social practice of fabrication. Glossed as temporary flaws en route to a fully up-graded state (that never quite arrives), these gaps cover for the selective, asymmetric, and obscure ways in which new standards of truth produc-tion affect different kinds of bodies. Sex, friendship, and happiness are datafied not in terms of what is most meaningful but what aspects of it can be rationalized at the lowest cost—and, increasingly, recombined and sold on for maximum profit. More broadly, the very turn to technology in the first place as a mode of organization, as a teleological harbinger of ob-jective truth, is a collective decision—however gradual and inadvertent—about what knowledge looks and sounds like in this society and what kinds of testimonies and experiences will be left to unofficial channels. The fantasies of the indefinite archive, of predictive control, of optimized humans, of the body as a computer all point to the emergence of the data-driven society's own generational set of systemic biases.

Default

Both state and self-surveillance exhibit a language of the "new normal" in which electronic dragnet surveillance, preemptively fabricated arrests, and smart machine–enabled data-sense are presented as a historical inevitability. To ask what ought to count as knowledge, is, in large part, to ask what it means for big data and smart machines to become our default and what it means to turn to technology as a default response to social problems.

In the shift from the Quantified Self (QS) to the Quantified Us, or counterterrorism's pressure to predict by every available evidence (how-ever circumspect), there emerges an unnerving norm that your truth must look and sound like data to be sayable and admissible. As the pro-cess of knowing oneself and making oneself known to others becomes automated and environmental, technologies of datafication act less as discrete tools and more as "paternalistic" organizers of human emotion and cognition.[2]

The political and moral stakes are most explicitly perceivable in the intersection of data and race, where technological solutions inherit a far longer tradition of asymmetric treatment. Efforts to resolve those historical biases—by mapping cases of police violence on African Americans, for example—run into the problem where alternative ways of truthtelling, embedded in local communities and personal experiences, are often more difficult to capture and represent. Thus, the standard response to the objections of marginalized populations: "We don't have the data. We don't know if what you're telling us is true."[3] The road to empowerment narrows into a set of technical requirements, reflecting broader differences in what kinds of databases are funded and which are not, what kinds of populations are heavily datafied and in what ways. In any case, becoming legible to databases does not guarantee fair and equal treatment. Facial recognition systems have been much criticized for their tendency to perform best with white faces and misrecognize colored ones. But is equality here achieved by "improving" the algorithm to better see black faces? Or, given the accumulated historical weight of racial biases, would it be fairer to delegitimize the use of such technologies for policing and carceral purposes?[4] For certain kinds of bodies, datafication might seem an empowering choice, a sovereign and individual decision to walk boldly towards the posthuman future. For others, to appear correctly in databases can be the unhappy obligation on which their lives depend.

The data-driven society thus retraces the steps taken by state administrations of high modernity, where cartography and other techniques sought to chart all of society, all of nature, in terms of state interests.[5] In seeking to comprehend populations and forests as resources for government, modern states produced a vast amount of unprecedented knowledge. Yet this also required a "narrowing of vision"; what the state sees is what the state may control, and thus, the citizen is required to become readable in a uniform and homogenous way.[6] Then as in now, progress is the hallowed name by which alternative ways of knowing are cleared away in favor of the chosen metric.[7]

To default, then, is also to narrow, to cut away other possibilities. The word has a curious mix of etymological roots. The vulgar Latin de + fallere and its Old French adaptations dance around concepts of fault, failure, lack, and even certain kinds of offenses that relate to a failure to take action—some of which survive in the modern use of "to default" to mean negligence of financial and legal obligations. At first glance, the defaults of technology involve a mundane and depoliticized sort of judgment; for instance, what should a piece of software present as an automatic, recommended option? What should appear first in a drop-down menu? Yet here, too, is a vestige of older meanings; a default entails an inability to think of alternatives, a narrowing of possibilities, and a situation wrought not out of explicit choice but of inertia and inaction.

Militating against such nuances is the myth of technological progress as a universalizing and unifying trajectory. The technological default is supposed to take humanity into its most optimal future, innovating and upgrading in a grand journey of civilizational improvement—as long as society accepts a way of seeing in which everything exists as a resource for extraction, calculation, and instrumentalization. Heidegger taught us long ago that the essence of technology is a way of knowing, a way of ordering the world. And "where this ordering holds sway, it drives out every other possibility of revealing."[8] The default to technology as a universal solution, one which engineers civilization progress (and is itself perfected through this irresistible trajectory), forms the groundless ground of the data-driven society.

A telling moment for this default can be found in a slightly earlier period of computational optimism. A decade before *Wired* editors Gary Wolf and Kevin Kelly founded QS, another network entrepreneur closely affiliated with the magazine, Nicholas Negroponte, proposed One Laptop Per Child (OLPC): a dream of bringing personal computing and the internet to the entire globe in the form of supercheap laptops. Latent in OLPC's lofty rhetoric was a magical short-circuit between the means and the ends. Give laptops to every child, it promised, and you will transform the world for the better. Asked how exactly the laptops were going to make a difference, Negroponte replied: nobody asked if it is really a good idea to provide everyone with electricity—the same goes for laptops.[9] We might well replace "laptops" with "data" or "artificial intelligence" to locate the generational reprisal of this groundless ground. Neither the NSA nor the Quantified Self has been quite so bullish. But the idea that big data and smart machines are "coming," and that society must make the best of it, forms the wider backdrop for discourses of data-sense and the new normal.

If the earlier enthusiasm around new media and personal computing coalesced around this trope of the computer as a universal good, the data-driven society enacts a similar mythologization of correlation. As society grows used to operationalizing correlations without ever discovering a coherent theory or causal relationship, the discovery of correlation, any correlation, starts to pass as the discovery of valid knowledge. Frank Pasquale has suggested that correlations discovered in databases should not immediately circulate as information but instead pass through a human audit that establishes external standards for what counts as a legitimate and meaningful kind of correlation.[10] As we have seen, the data market expands by taking every technique of datafication and asking, "Could we scale this up? Could we use this everywhere else?" Technology start-ups constantly define themselves in these terms: a TikTok for the elderly, an Uber for bikes, a Fitbit for your brain. After all, it is precisely this frictionless and cross-contextual control creep that allows datafication to extend capitalism's search for surplus value. For all the moral and philosophical reflection that accompanies new technologies, their fate tends to be decided primarily by two thresholds: Is it technologically possible? And will it turn a profit?[11] Yet this model of omnivorous technologies yields precisely the epistemological tyranny that privileges whatever looks and sounds like data over everything that does not.

Such defaults of thinking and feeling about technology amount to what I described as a general "image" of technoscience, a broad and malleable faith that technology will come good. Laptops will lead to a better world somehow, even if the details are all so complicated. And thanks to the gestures of deferral built into these justifications—just wait for the next version, the next project, the next invention—it is a flexible kind of faith that survives the unfulfilled promises that pile ever higher by the day. The default to technology thus engages in its own form of futures trading, a constant trade of "promises as currency"[12] that siphons public faith from one imperfect vision to another. Even as we constantly debate whether all we have done in the name of progress was worth it, it is precisely this incompleteness of the project that invites us to keep dreaming.[13]

This is not to say that tech optimists believe the consequences of technological infrastructures to be perfectly uniform and predictable. What carries the day is the general belief that even if technology is not always for the best, this optimism must be *true enough* that we might not inquire further. Sheila Jasanoff has argued that we continue to mystify technology through ideas such as technological determinism and unintended consequences, depicting it as a force that is fundamentally just as unimaginable as it is inevitable—and therefore beyond the reach of ethical thought.[14] We might recall the refrain of the QS community: its founders do not say that "we chose to pursue this technology on behalf of everybody." Once unlocked at a technical level, the smart machines are depicted as already on their way, the only choice being whether to ride the waves or flee from them. In this kind of thinking, emergent implications, such as the co-option of fitness tracking by insurance companies, are regrettable accidents that could not have been predicted and therefore cannot be anyone's specific responsibility. Even as technologies of datafication are sold as remarkable machines of prediction and control, any responsibility for the consequences of these technologies is excused and passed off into the caprice of historical chance.

Technology thus constitutes not just a set of machines and functions, or a set of instrumental attitudes, but—recalling Weber—a moral cage. Concepts of efficiency and optimization, progress, and innovation comprise a style of reasoning that consistently protects technology from judgment by external standards. Technology is asked only to continue to instrumentalize the world, to help us to do anything and everything faster, cheaper, and at larger scales. Genuine moral reflection gets squeezed out: the values that produced modern technology in the first place are not going to betray it in a debate about how we should live. Yet the cage itself remains imperfect. Since the earliest days of computing technologies, the specter of the Manhattan Project has loomed large over every discussion of technological morality. The atomic bomb is a technology which can be used, but ought not to be; standing—perhaps temporarily—as a deterrent to the narrative of ineluctable progress, it has compelled the question of whether technological change is inevitable and whether it is a good in and of itself.[15]

To problematize the default of technology, then, is to repeat the counsel of Joseph Weizenbaum: not everything that *can* be done with technology ought to be done. This correction is especially necessary in light of the many ways in which technology is justified through future promises. As human subjects, who will always find themselves in the incompleteness of the present, we need better ways to account for the actualized distribution of those benefits and harms.[16] If datafication empowers individuals with better knowledge, does it also empower institutions? Employers? Authoritarian governments? Is it not possible that some technologies must be rejected or constrained *despite* their many benefits?

Critique

The enduring image of technology as a universal and progressive force proves deeply restrictive for our efforts to engage in meaningful critique. In repeated honeymoons with the fantasy of machinic objectivity, society pays up front for the brilliance of unfulfilled dreams with the "side effects" of present interventions. Importantly, the position of the critical subject itself is shaken by what we have said about the actionability of knowledge, the bypassing of human cognition, and the systematic dependency on uncertainties and simulations. For a long time, the look and feel of critique has involved a "hermeneutics of suspicion":[17] it begins with a leak, an exposé, a revelation that ironically resembles the pseudo-journalistic bait in the attention-driven digital economy: "you'll never believe what your smartphone does behind your back!" The new information is meant to equip the subject with a fresh awareness of the political struggle that already defines their social existence and thereby empowers them to take action. This model of critique is itself grounded in the mythical figure of the good liberal subject, the agent who retains some ability to engage in purposeful and individual resistance against structures of control.

Yet it is exactly this figure that is being disrupted and bypassed by technologies of datafication. Data-driven knowledge is sold as a slick and smooth "insight," murky and opaque to human cognition. At the same time, its enormous scope and complexity require an ever-growing panoply of mediations to allow human subjects to cope. (Historically, of course, we may trace this double-play of concealment back to computers as a whole.[18]) The space of moral response—of hesitant explorations, of little transgressions, of waiting and reflecting, of cobbling together a different order of things—is optimized away in favor of ever more efficient calculation. As Wendy Chun said, we are kept busy updating to stay the same.[19] The mantra of big data is that all we need is a correlation that can be manipulated, not causality or theory. The corollary is that we, the human subjects, are told that we do not need to know why; we need only implement the recommendations of the algorithm, and we will be better off—even if no wiser. We will be fitter, happier, and more productive without necessarily understanding what it means to seek those things or to define and aspire to them in the particular way that we do.

These dilemmas call into question the status of knowledge itself as an unquestioned good. Technologies of datafication expand under the banner of better knowledge. In the process, they introduce new forms of speculative and deferred truth production, externalizing the truth of the body, the choices we make, and the freedoms we enjoy out toward a distributed network of machines and corporations, databases and state institutions. The corollary is that "better knowledge" is not necessarily better for the human subjects involved and that forms of invisibility, disengagement, and underdetermination need to be defended through more robust moral and political scaffolding.

The dangers of data-driven knowledge are widely recognized. Construed in terms of issues such as data privacy, corporate responsibility, and transparent governance, mechanisms of datafication have increasingly prompted major regulatory initiatives. Yet this form of problemsolving itself often relies on a liberal grounding, one that seeks to restore and protect the sovereignty of the informed subject. In this picture, knowledge *about* individuals is treated as possessions to be protected: one must own one's data or, at least, provide informed consent to its usage. Transparency and public education initiatives enjoin individuals to ingest more and more knowledge to properly exercise their choice over the conditions of their datafication. Meanwhile, the good liberal subject—who "knows" for themselves—is consistently undermined by the privileging of machinic sensibility, of indefinite archives, and of predictive control. Efforts to prop up this model by adding more layers of technological and regulatory support is helpful in the short term but risks a certain neoliberal divestment. The work of knowing is increasingly laid on the shoulders of overloaded individuals, even as the structural conditions of new technologies withdraw their support for a traditional kind of informed public.

In this light, we might ask: What if "better knowledge" should not be the primary goal of our information technologies? This is certainly not to ask for the reign of ignorance and secrecy or some atavistic disavowal of the Enlightenment. Rather, the point is to recognize the consequences of fetishizing knowledge as a social good. Recalling the value of nonac-knowledgment, the dangers of transparency, the dilemma of "bypassing" human sensibility, we might ask: When should we resist the temptation to datafy and predict, to produce knowledge out of human bodies and lives? How can we temper the commercially attractive pursuit of datafi-cation to allow more clear-eyed acceptance that much of what we know remains uncertain? The ambitions of datafication demand the most stringent standards for what counts as knowledge and how we should use imperfect data. The many injustices and asymmetries of datafication arise not because we do not *yet* have all the information but because we have too much information that we utilize in indisciplined and inap-propriate ways.

None of this is to sound the death knell for liberal values, for the public use of reason, and for individual freedoms. Although mediated and partial, the degrees of choice liberal democracies have enabled over one's political leaders, one's religion, and one's way of life are precious rights that are under renewed threat. These values need help, not cyni-cism. What should concern us is the uses to which technologies of data-fication are being put, driven by the interests of the data market, the political exigencies of the war on terror, the misguided idealization of machinic objectivity. Too often, data-driven surveillance is being used not to empower us for the choices that matter but to occupy us with what does not: the ballooning of speculative factmaking, which muddies the waters of public knowledge and debate, and the many little choices, the many exposes and scandals, that the fetishisation of transparency and optimization would have us spend our every waking hour on. As the way we know and the way we put that knowledge into action is increas-ingly externalized, what is required are new ideals of representation and visibility—ones that protect some degree of sovereignty over the truth of who I am, what kind of life I shall lead, and how those choices will be made to count against the datafied judgments of society and the state.

What does meaningful critique look like under such conditions? The search for an effective and systematic moral criticism of technology is underway across several different scholarly traditions. Yet the question remains as to how to deal with the changes in epistemic thresholds and the erosion of the integrity of the liberal subject. For instance, a great deal of moral philosophy vis-à-vis new technologies builds on utilitarian foundations and attempts to sort out the moral status of technological objects and of technologically affected persons. Is an augmented human deserving of higher moral status than humans? Who, or what, is responsible when a (quasi-)autonomous technological object such as a self-driving vehicle kills a human being? How might such responsibilities be allocated across human, nonhuman, and quasi-human entities? From bioethics to machine ethics, such analyses are focused on areas where ostensibly traditional definitions of human rights and responsibilities are being disrupted.[20]

In other words, we again find a certain liberal conception of an autonomous human subject as the presumed norm, enjoying full privileges conferred in the name of human rights or ethical behavior, an "individualist ontology."[21] It is against this figure that the rights of robots, posthumans, and brains in a vat may be calculated and distributed. In a way, the very fact that such inquiries must now focus on "intelligent" machines or augmented humans speaks to the ongoing breakdown of the boundary between humans and things and the erosion of a singular standard for what qualifies as a "legitimate moral agent."[22] Individualist ontology seeks to repair, or at least adapt, the Kantian imperative: treat people as ends in themselves, while that which we treat as means, we call things.[23] The separation of people and things, and the effort to preserve this separation by carving out many hybrid subcategories of thinglike people and peoplelike things, permits a relational model in which humans (or, rather, whatever qualifies fully or partly as "humans") may be served by things through the instrumentalizing power of technology. When technological objects start to achieve greater autonomy (in multiple ways) and serve priorities and ends not so tightly tethered to the user at hand, this moral separation of means and ends runs into trouble.

In such a schema, the relative positions of these ethical agents are used to calculate the distribution of harm. Harm itself is most conveniently conceptualized as a discrete and actualized sort of injury—and less conveniently as forms of potential, atmospheric, shared, structural violence. This book has shown how American juridico-legal systems try to cope with the idea of electronic state surveillance: Could you, the ostensible victims, please prove that you have been subject to individualized and particularized harm? Such harms are, in turn, counted through the apparatus of risk, which codifies the distribution of damages across

ethical agents in calculative terms. Yet how amenable to such codification are situations such as the automation of self-knowledge or the
environmental extraction of personal information by states and corporations? It is increasingly questionable as to how well such moral codes
can handle uncertainty, including the kinds posed by the complexity of
new technological systems.[24] To be sure, newer models for risk calculation such as the precautionary principle (arguably dominant since the
later twentieth century) have sought to become more flexible, building
in various buffers for uncertainties and irrationalities.[25] But such systems are fundamentally based on the reduction of human subjects' felt
uncertainties into probabilistic values. Sven Hansson calls it the tuxedo
fallacy: although oftentimes we tread in jungles, full of unknown and incalculable dangers, we dress for the casino, that comfortingly closed system of known variables.[26] This strategy of formalizing the consequences
of technology into calculable risk factors, and deriving a probabilistic
schema for optimal outcomes, provides a highly specialized instrument
with its own assumed grounding.[27] What kind of judgment, consensus,
and principle underwrites such moral codes in a society where human
decisions are increasingly shaped a priori through machinic ones? What
kind of imagined default for a good society is at work when we tie ethics
to the mathematical problem of minimizing risks and harms?

Of course, in the ebb and flow of the everyday, the little decisions,
compromises, nudges, we have always relied on a host of non-calculative
compasses. Shannon Vallor has proposed that we return to the venerable tradition of virtue ethics, delving into Aristotelian, Confucian, and
Buddhist traditions to navigate the latest technological dilemmas.[28] The
crucial point here is that virtues transpose the center of ethics from the
process of decision to the human subject. It does not and cannot pre-
scribe any consistent formula on how to arbitrate any situation; instead,
it calls for the kinds of virtuous subjects who may be trusted to make
particular and even exceptional judgments. This is not a failure of virtue
ethics; rather, it reminds us that there is no "single best way to live" in the
data-driven society and that knowledge, choice, and responsibility are
not things that belong discretely to a rule or an individual. If there is no
optimal point toward which ethics can strive, then surely our algorithms
cannot capture the world in any such singular way, either. These explorations emphasize the need for moral standards that are external and inde-

pendent, that is, ways of discerning and deciding that speak a different language to that of technological rationality. By systematically refusing the idea that moral decisions can be codified and depersonalized, virtue ethics seeks to escape the moral cage of technological thinking.

Another direction for disrupting the cage would involve diversifying the data itself, loosening the hold of quantitative and correlational tendencies on processes of factmaking. Nora Bateson has pushed for the development of "warm data": contextually and relationally rich forms of information that might be more accommodating of human contradictions and inconsistencies.[29] It would be a kind of data less amenable to computational logic than the "cooler" sort but render our knowledge of complex systems (including humans) more adequately. But such data is by definition less legible, calculable, manipulable, and profitable in the eyes of digital technological systems that are founded on the reduction of such warmth to begin with. Can warm data retain its relevance when push comes to shove to predict, to optimize, to decide? How can knowledge that is not easily datafied—the affects, the interrelations, the lived experiences, the tacit knowledge, and so forth—hold its own in a world where we learn knowledge is what looks and feels like (cooler) data?

Attempts to enforce such external moral standards must account for the ways in which technologies of datafication have transformed the would-be moral subject. The focus on virtues returns us once again to the chimerical figure of the subject. In place of precise rules and algorithms that anyone might employ to reach the right decision, the virtuous human individual must carry the burden of knowing how to act. Vallor describes virtues like "prudential judgment" and "appropriate extension of moral concern" for a data-driven society.[30] But these are long-standing values that we have not exactly forgotten. The crucial question is how new technologies prefigure human experience and human judgment such that we end up straying from those virtues. As Latour said, we should not be asking what it is about drivers that make them speed in school zones but how solutions such as speed bumps inculcate a (partly, at least) nonconscious disposition toward responsible driving.[31] How might we conduct critique, and build a way of thinking morally about technology, in a way that does not fall back on the assumed figure of the liberal subject, of individual choice as the atomic unit of ethical behavior, of "knowledge" as a linear accumulation of information that enables better such choices?

Politics

One answer is that critique must deliberately disrespect the rationality of technology. Data-driven truth production invariably steers us toward a greater dependency on smart machines and data-driven reasoning to establish our picture of the world. The moral cage of technological rationality hinders every effort to think of technology as more than its functions, and to hold them to account by terms other than its own. Although guns may not quite kill people without people behind them to pull the trigger, it becomes morally necessary to act on that technology to address the social problem—without leaving it up to some mythico-capitalistic faith in the fairness of the market. The question is what kind of broader political and moral debates are being excised when we end up talking in terms of efficiency, of optimization, of using data to ascend the pyramid of knowledge.

This "disrespect" means stepping beyond the role of pointing out technology's present abuses and flaws. To return to an earlier example, insisting that facial recognition algorithms improve their capture of black faces renders society further dependent on those algorithms.[32] Data will always travel across contexts, and it will always beget conditions of use that hybridize with all the existing flaws and biases of our society. Technology is not reducible to simply means or ends, for it produces its own unexpected ends precisely by being the means of things.[33] It cannot be an issue of how to design guns such that they will only shoot bad people. The question lies outside technology: What kind of society do we build and export and leave behind such that it furnishes a discursive, economic, and material context for this or that use of guns?

In other words, we must ask *political* questions of technology—even as technology often fiercely claims itself to be apolitical. The assertion remains just as obvious, and just as necessary, as when Langdon Winner posed it in 1980: "there is no idea more provocative than the notion that technical things have political qualities."[34] As we have mentioned, a crucial aspect of the fantasy of epistemic purity is the desire that we can change the world without engaging with the swamp of politics: that by extracting new and wondrous kinds of raw data and processing them through groundbreaking new mechanical systems, we might simply erase a formerly intractable social problem from existence. *Silicon Valley*, the

American comedy show dedicated to this enduring stereotype of the techie, is full of endearingly tunnel-visioned geeks too nerdy for politics or ideology. At the same time, there is an earnest belief that the work they do will create a better society . . . but is that, in fact, not the goal of politics?

The physical environment of *Silicon Valley* itself is a kind of silo: it is a place populated by overdesigned mega-campuses and scooter-riding coders, suburban homes overtaken by open circuit boards and unwashed geek geniuses. They are seemingly entirely insulated from the rest of the world and its messy problems by a layer of venture capitalist funding. But this is not quite where technology is born; every invention, every device has a far more global footprint that this depoliticized imagination leaves invisible. In 2009, Sun Danyong killed himself at a Foxconn factory, one of the key Chinese sites for the production of Apple iPhones. The next year, at least eighteen more would do the same. For a time, the suicides brought global attention to the direct relationship between the latest technological marvel for the developed world and the problems of global economic exploitation. It was one of many reminders that technology is no separate industry or culture but a certain way of knowing and doing that is woven into everything else. Yet the fiction persists that these are mere episodes of abuse, side effects that are somehow secondary to the core effects of technology. Over time, the world would largely forget Sun Danyong and the blood on the phones. The suicides would continue.

This illusory separation is well served by the aesthetics of big data and smart machines, which consistently fantasize about the dematerialization of technologies. During the mid-2010s, the catchphrase "data is the new oil" gained traction. Data was described as the next frontier of surplus value, something to be discovered, claimed, extracted, and monetized. But what is more revealing is the aspects of data—and oil—that are left out of this telling. Data may be the new oil, but the messy and toxic conditions of its extraction and processing are elided in the ethereal tropes of the cloud.[35] Everything is there at your fingertips; the storage is infinite; the delivery, instant. State surveillance dragnets do not flout their presence but recede from human experience at every opportunity, even sinking into tapped undersea fiberoptic cables at the bottom of the Atlantic Ocean. Self-surveillance tools dream of "frictionless" monitoring, where user discomfort of (and resistance to) new products is minimized as the machines blend into the background of everyday life.³⁶ Such designs extend the fantasies of older generations of media. Electric communications media of the nineteenth century too had promised to eliminate time and space—a fantasy that served a highly politicized imagination of a homogenized and instant world in which difference is mere exotic distraction and convenience is our eternal reward.³⁷ In the data-driven society, the depoliticization of technology combines recycled ambitions of a brilliant future with a certain political and social conservatism: everything can and should stay the same, just faster, cheaper, and easier.³⁸ No surprises, only upgrades.

Not for Us

Specifically, the advent of big data and smart machines brings to the forefront the sheer alienness of technological rationality. In learning to experience one's own eating, running, and sleeping in relation to the quantified reports and recommendations of tracking devices or in adapting one's behavior to stay clear of the predictive algorithms that might flag us for preemptive arrest, the human subject encounters data-driven knowledge as a foreign and autonomous mechanism that increasingly compels adaptive acquiescence. People are not asked to become more like machines; the machines and the knowledge they produce are too complex, too different, for that. Instead, we are asked to cooperate with the work of the machines that we do not understand, rearranging the way we think and the way we make decisions to become more legible to the epistemic priorities of smart machines. In short, any moral and political scrutiny of datafication must begin from the point that these machines and the "better knowledge" they produce are, increasingly, *not for us*.

To say technologies of datafication are not for us is to say that society is increasingly reorganized in terms of those technological priorities. Consider, for instance, the value of privacy. It is declared dead and buried every other week in the data-driven society and yet remains a dominant frame for assessing those technological changes. It is well understood that the rationality of big data is inherently biased against any traditional provision of an individualistic, "my house is my castle" form of privacy. From the injunction to share on social media platforms to loyalty card programs at the local supermarket, datafication operates according to its own (often commercial) priorities for maximizing returns from human subjects—whether by selling users' data on to third-party clients or leveraging it to optimize their business operations. Asking governments and corporations to voluntarily respect individuals' privacy is effectively demanding that technology sabotage its own efficiency. We must begin from the recognition that the default of datafication is to eliminate privacy wherever it goes, indifferent to any liberal humanistic concerns. In everyday parlance, the user is someone to be served and serviced, someone who actively makes use of technologies. In terms of technological priorities and rationalities, the user is someone "about whom information is recorded and processed," someone who can be mined and processed for profit.[39]

Overtaken by such technological priorities, privacy becomes less a question of "Who knows what I did last night?" and more a graduated yardstick for the degrees of recessivity, degrees of distance, between human subjects and the technologies that surround them. (This is, of course, less a historical erosion of individual privacy than a growing realization that modern privacy has always been a porous, flexible, and publicly contested value.[40]) In other words, to talk privacy in the data-driven society is to not demand a hermetic seal between private and public but to demand more humanly meaningful control over the ways in which bodies are datafied and manipulated at a distance. If we are constantly categorized, predicted, judged, and condemned beyond the horizons of our experience, then "privacy" might now involve mitigating the recessivity of data-driven knowledge. What is at stake is not (only) the right to be let alone but our ability to understand, navigate, and perhaps even affect, the ways in which we are seen and judged and measured. In an oft-told story, Samuel Warren and Louis Brandeis are said to have concocted the traditional concept of privacy in response to a paparazzi intrusion of a high-society wedding. The sentiment was clear: this unregulated leakage of the formerly private could not be tolerated. Today, it is no longer a question of stopping the leaks but of what kind of control, and freedom, can be secured within the necessity of circulation.

On the other side of the computer screen, what machines do beyond our supervision and our understanding must also be assessed vis-à-vis the alienness of technological rationality. Automation is advertised as an instrument of convenience and efficiency—a classic bedtime story in which social relations remain identical even as they become faster, easier, and cleaner. Rather, it should be understood as an exercise of political redistribution: a shift in which the inscription, circulation, and even interpretation of information are increasingly designed not for human users but for smart machines. If the original uncanny valley was coined in the 1970s to describe machines' attempt and subsequent failure to align themselves to human sensibilities, smart machines communicate their "not-for-us"-ness in their everyday uncanniness: the automatic door that refuses to open, the fingerprint recognition system that takes a dozen tries, or, as one media theorist's young daughter complained, the "magic potty" that creeps out young children with its ill-timed flushes:[41]

> So many ordinary objects and experiences have become technologised—made dependent on computers, sensors, and other apparatuses meant to improve them—that they have also ceased to work in their usual manner. It's common to think of such defects as matters of bad design. That's true, in part. But technology is also more precarious than it once was. Unstable, and unpredictable. At least from the perspective of human users. From the vantage point of technology, if it can be said to have a vantage point, it's evolving separately from human use.

We might also say that it is us humans who grow more uncanny by the day, contorting our behavior to align ourselves, however awkwardly, with the machines that will go of their own accord. To develop datasense is to learn to flash a trained smile at the phone to unlock it in the middle of a heated argument, to look just right for the facial recognition system one's employers have hired to make sure everybody smiles the right way for the customers,[42] to sit just so for the whims of the magic potty. As the Quantified Self becomes the Quantified Us and as electronic surveillance systems seek an ever-wider range of preemptive indicators for crime, we must expect to confront new ways in which people are asked to see like machines and to become visible to machines. This becomes the daily task whenever we seek to produce truth about ourselves, truths that *matter*—for getting jobs, for navigating racial bias, for learning to use new technologies, for political representation, and for staying on the right side of the ever-growing databases and algorithms that mark bodies out for special treatment, police attention, social erasure, death, and violence.

The cruel joke is that as this "alien" technological rationality grows indifferent to human meaning, what was supposed to be the primary purpose of data-driven knowledge—to enable and justify wiser action—begins to corrode. In a world where so much of what we call knowledge is unreadable for human subjects, data provides not so much a full understanding that is a basis of considered opinion but a scattershot array of instructions, orders, insights, and predictions. Something similar had happened in the mid-nineteenth century, when an earlier generation of statistical knowledge production began to proliferate. As Ian Hacking recounts, the rapid adaptation of statistics and probability to social problems provoked a popular backlash: all these numbers are well and good, and they certainly give institutions and governments a guide for action, but what am *I* to do?[43] If this is the normal distribution of mortality, if this is the demographically specific probability of disease—well, what am I, the individual, meant to decide and do about my still unknowable fate? Similarly, data's promise of better knowledge aspires to a kind of truth that communicates better with nonconscious tendencies of the human psyche than to its conscious subjects, better with patterns, correlations, and predictions than moral exigency or irregular and nonpatterned activity, better with tech-literate lives already organized around certain ideal images of the flexible self-optimizing laborer than the many other kinds of living that still define the human.

In optimists' projections, automation promises an immediate and finalized presentation of fact: just press the button, or call out to the machine, and it will deliver the desired end product without fuss. Some technical objects might still demand a degree of lived engagement, a process of learning and cooperating through which humans find themselves transformed. But what is fantasized in promises of "frictionless" and backgrounded tracking is often not so much the empowered freedom of the liberal subject but a devalued accumulation of things and facts divorced from meaningful experience that might genuinely equip human subjects for better decisions.[44] In short, the question of what *ought* to be done through technology is a question of what we can afford to leave opaque and distant to human sensibility, of what we should and should not be pushing out of our own horizons and up to the interests of technological rationality.

* * *

When we consider that technology is "not for us" and that the fabrication of numbers and correlations is not our inevitable path to truth, we come away with a different sense of the ought. The populational explosion of smart machines presents us with a political problem of access, priority, and communication between humans and technological objects. This is precisely why we must turn away from the mistake of thinking that the technological *can* equates to a human *ought*. A century ago, Simone de Beauvoir wrote that science fails if it is consumed by the quest to attain and capture being. Rather, "it finds its truth [as] free engagement of thought," a pursuit of possibilities. And so "technics itself is not objectively justified"; its focus on convenience, efficiency, and luxury amounts to an obsession with "saving up existence" when the focus should be on how to expend it wisely.[45]

We began with a question: What if that figure of the human subject, always historical, always precarious, always long in the shadow of its own myths, is being swept away in the data-driven society? To be sure, its passing looks less like the advent of thinking robots that replace us and enslave us (a human-centric fantasy itself, wherein a superior kind of human lives on in our stead), and more a slow erosion. It is the redrawing of the lines that sought to secure the idea of an autonomous individual, of humans who process facts and make reasoned judgment, of humans who might determine their own conditions of social existence. The data-driven society is not defined, at least not yet, by a binary division of threatening or superior machines and the dethroned humans. Neither is it the ascendance of humanity writ large toward a fantastic posthumanism. As it has been for so long, so it is again a question of what counts as being human, of how life becomes fitted into calculable categories of the normal, the monstrous, the dangerous, the optimal. Neither the uncanny tricks of smart machines nor the sublime complexity of big data should distract us from this moral problem.

ACKNOWLEDGMENTS

This book is about values: the subservience of human values to the rationality of technology and capital and, specifically, the transmutation of knowledge from a human virtue to raw material for predictive control. Working on such a project has, over the years, raised another question: What is the value of ideas, of philosophy, of a single book, in such a society? The degree to which this book has any answer owes much to the generous criticism of many kind souls.

It all began with a downtown bookstore in Auckland, New Zealand, which had made the economically mysterious decision to display copies of Michel Foucault's *The Order of Things*. The book, which remained completely inscrutable to me for a long time, nevertheless helped me glimpse a language and world beyond my own understanding. Years later, on my first day at the University of Pennsylvania, Carolyn Marvin suggested that I consult a certain book called *A Thousand Plateaus*. For three months, I tread back and forth in that maze, aware that Deleuze, in particular, was laughing from his grave. Certain ideas, it seemed, were worth getting lost for, their value lying beyond immediate use and in the years to come.

The effort to capture something of that value, and re-create it, has again required the help of many learned folk. Tony Schirato graciously explained that my writing was too "gothic" and that true mastery of complex ideas is the ability to express it simply. Tony, may he rest in peace, would have been amused to know that I've taken the lesson to heart—though without too much success. Stephen Epstein generously supported an early research project that was not particularly interesting, helping me eventually arrive at the themes of this book. Sharrona Pearl and Marwan Kraidy joined Carolyn Marvin in reading through the dissertation, offering valuable feedback and support on what was the least readable version of this manuscript. Jim Paradis and William Urrichio graciously entertained my streams of thought while at MIT, while Eric Zinner, Dolma Ombadykow, Martin Coleman, and others at NYU Press have ensured that the book is presentable to decent folk. The anonymous reviewers read the manuscript with such care that they were able to produce a truly wide-ranging list of its flaws—and for this, I am immensely grateful.

The ideas and stories in this book gestated through a number of places. The first drafts began in Amsterdam, where I was mentored by José van Dijck and hosted by Joost Vecht, who generously allowed me to stay in his spare room full of a still unknown, possibly mildly toxic, turquoise powder. Work continued at the University of Pennsylvania, in conversation with colleagues including Sandra Ristovska, Piotr Szpunar, Yoel Roth, Bo Mai, and Aaron Shapiro. The stories in the book were refined through presentations and discussions at MIT; Microsoft Research New England; Internationale Kolleg für Kulturtechnikforschung und Medienphilosophie at Bauhaus-Universität Weimar; University of California, San Diego; the McLuhan Center at the University of Toronto; the Wattis Institute for Contemporary Arts; Copenhagen Business School; University of Milano-Bicocca; and more.

Help arrived in many different forms. Most of my friends have wisely refrained from asking what exactly this book is supposed to be about and what is taking so long—except for Jin Woo Kim, who spent several years suggesting a career change to a failed K-pop artist. Two cafés—Harvard Square's Crema and Vancouver's Honolulu—unknowingly supported the writing of most of the manuscript. Most importantly, I would like to thank my family: all three parents, two cats, and the one and only Minji, from whom I have received far more than a book can say. The result is a book that attempts to speak about human values, to help us navigate the shifting ground beneath our feet in the age of big data and smart machines. It is my hope that intelligent readers will find a few useful tools to take away from an imperfect, if well-intentioned, work.

NOTES

INTRODUCTION

1 Daniel W. Smith, "Deleuze and the Question of Desire: Toward an Immanent Theory of Ethics," *Parrhesia* 2 (2007): 66–78.

2 For instance, the study of epistemic cultures in science examines "how we know what we know" and what kinds of "epistemic machinery" furnish the actual claims (such as equations or laws). Karin Knorr Cetina, *Epistemic Cultures—How the Sciences Make Knowledge* (Cambridge, MA: Harvard University Press, 1999). I also draw on Ian Hacking's notion of styles of reasoning later.

3 Joseph Weizenbaum, *Computer Power and Human Reason: From Judgment to Calculation* (New York: W. H. Freeman and Company, 1976).

4 Mary L. Gray and Siddharth Suri, *Ghost Work: How to Stop Silicon Valley from Building a New Global Underclass* (Boston: Houghton Mifflin Harcourt, 2019).

5 For examples in the twentieth century, see Colin Koopman, *How We Became Our Data: A Genealogy of the Information Person* (Chicago: Chicago University Press, 2019); Orit Halpern, *Beautiful Data: A History of Vision and Reason since 1945* (Durham, NC: Duke University Press, 2014); Paul Erickson, Judy L. Klein, Lorraine Daston, Rebecca Lemov, Thomas Sturm, and Michael D. Gorin, *How Reason Almost Lost Its Mind* (Chicago: University of Chicago Press, 2013).

6 Long used as a generic term for police techniques such as stop-and-frisk, the dragnet became a popular label for key elements of NSA electronic surveillance programs unveiled by Snowden. For instance, the NSA's upstream interception of phone call records from "backbone chokepoints" of the communications infrastructure is described as a dragnet in lawsuits brought by civil society against the NSA (*Clapper v. Amnesty International* [2013]; *Wikimedia Foundation v. NSA* [2015]). There is, of course, no universal dragnet for the complete capture of all communications data. Rather, the dragnet expresses the relative jump in the size and breadth of the NSA's numerous collection programs and the many ways in which the data captured necessarily precedes and exceeds the identification of specific targets. Also see Benjamin Wittes, "The Problem at the Heart of the NSA Disputes: Legal Density," *Lawfare*, February 14, 2014, https://www.lawfareblog.com/problem-heart-nsa-disputes-legal-density.

7 Bruce Neuman, "When a Senator and the C.I.A. Clash," *The New York Times*, March 12, 2014, https://www.nytimes.com/2014/03/13/opinion/when-a-senator-and-the-cia-clash.html.

8 Respectively, Tom Engelhardt, *Shadow Government: Surveillance, Secret Wars, and a Global Security State in a Single-Superpower World* (Chicago: Haymarket Books, 2014); Julia Angwin, *Dragnet Nation: A Quest for Privacy, Security, and Freedom in a World of Relentless Surveillance* (New York: Times Books, 2014); Greenwald, *No Place to Hide: Edward Snowden, the NSA and the Surveillance State* (London: Penguin Books, 2014).

9 Ariel Garten, "Know Thyself, with a Brain Scanner," filmed at TEDxToronto, Toronto, Canada, September 2011, video, 14:49, www.ted.com.

10 This included the full population of media coverage on self-surveillance and the QS from a selection of major publications—*The Atlantic, Fast Company, Harvard Business Review, Inc., National Review, The New York Times, The Washington Post*, and *Wired*—between 2007 and 2015. These were supplemented by coverage from other news sources; the QS official website archives; promotional material on commercial self-surveillance products, from product websites to crowdfunding campaigns to scientific research papers published by developers.

11 This approach draws on Funtowicz and Ravetz's analysis of what kinds of uncertainties are integrated into or excised from the normative bandwidth of scientific inquiry. See Silvio O. Funtowicz and Jerome R. Ravetz, "The Emergence of Post-Normal Science," in *Science, Politics and Morality: Scientific Uncertainty and Decision Making*, ed. Rene von Schomberg (Springer-Science+Business Media, B.V., 1993), 85–123.

12 For a systematic description of big data in these terms, see Rob Kitchin, "Big Data, New Epistemologies and Paradigm Shifts," *Big Data & Society* 1, no. 1 (2014), https://www.theoryculturesociety.org/kittler-on-the-nsa/; Rob Kitchin and Gavin McArdle, "What Makes Big Data, Big Data? Exploring the Ontological Characteristics of 26 Datasets," *Big Data & Society* 3, no. 1 (2016): 1–10.

13 Mei-po Kwan, "Algorithmic Geographies: Big Data, Algorithmic Uncertainty, and the Production of Geographic Knowledge," *Annals of the American Association of Geographers* 106, no. 2 (2016): 274–282.

14 Consider, for instance, Steven Shapin's history of Mertonian moral equivalence and other reshapings of scientific subjectivity vis-à-vis the increasing institutionalization of "Big Science." Steven Shapin, *The Scientific Life: A Moral History of a Late Modern Vocation* (Chicago: University of Chicago Press, 2008).

15 Michel Foucault, *The Order of Things: An Archaeology of the Human Sciences* (London: Routledge, 2002).

CHAPTER 1. HONEYMOON OBJECTIVITY

1 "Meet Sense," Hello.is, accessed April 13, 2016, https://hello.is/videos#meet-sense.

2 James Proud, "Goodbye, Hello," *Medium*, June 12, 2017, https://medium.com/@hello/goodbye-hello-c62ea1f58d13.

3 Fahrad Manjoo, "Mysteries of Sleep Lie Unsolved," *The New York Times*, February 24, 2015, http://www.nytimes.com/2015/02/26/technology/personaltech/despite-the-promise-of-technology-the-mysteries-of-sleep-lie-unsolved.html?_r=0.

4   Genevieve Bell and Paul Dourish, "Yesterday's Tomorrows: Notes on Ubiquitous Computing's Dominant Vision," *Personal and Ubiquitous Computing* 11, no. 2 (2007): 133–143.

5   David Haskin, "Don't Believe the Hype: The 21 Biggest Technology Flops," *Computerworld*, April 4, 2007, https://www.computerworld.com/article/2543763/ computer-hardware/don-t-believe-the-hype--the-21-biggest-technology-flops. html.

6   Jackie Fenn and Hung LeHong, "Hype Cycle for Emerging Technologies, 2011," Gartner, July 28, 2011, https://www.gartner.com/doc/1754719/hype-cycle-emerging-technologies.

7   For example, John M. Jakicic, Kelliann K. Davis, and Renee J. Rogers, "Effect of Wearable Technology Combined With a Lifestyle Intervention on Long-Term Weight Loss," *JAMA: The Journal of the American Medicine Association* 316, no. 11 (2016): 1161–1171.

8   Eric Kluitenberg, "On the Archaeology of Imaginary Media," in *Media Archaeology: Approaches, Applications, and Implications*, ed. Erkki Huhtamo and Jussi Parikka (Berkeley: University of California Press, 2011), 48–69.

9   A point well made in anthropological investigations of ritual. See Roy Rappaport, *Ritual and Religion in the Making of Humanity* (Cambridge: Cambridge University Press, 1999); Victor Turner, "Liminal to Liminoid," in *Play, Flow, Ritual: An Essay in Comparative Symbology*, ed. Victor Turner (New York: Performing Arts Journal Publishing, 1982), 53–92.

10 Benjamin J. Muller, "Securing the Political Imagination: Popular Culture, the Security Dispositif and the Biometric State," *Security Dialogue* 39, no. 2–3 (2008): 211.

11 Suzanne L. Thomas, Dawn Nafus, and Jamie Sherman, "Algorithms as Fetish: Faith and Possibility in Algorithmic Work," *Big Data & Society* 5, no. 1 (2018).

12 See the definition of fantasy given in Lauren Berlant, *Cruel Optimism* (Durham, NC: Duke University Press, 2011), 2.

13 Here I am drawing on existing arguments by historians of science and technology regarding the historical emergence of technology as word and idea, its expansion from specific crafts and solutions into a broader and autonomous force, the gradual prioritization of progress as for its own sake, an originary link between the focus on "rational manipulation of the environment" and the penchant for technological solutions, and America's historical depiction of technology through, paradoxically, often aesthetic and fantastical imagery of rational mastery. For examples, see Leo Marx, "Technology: The Emergence of a Hazardous Concept," *Technology and Culture* 51, no. 3 (2010): 561–577; Langdon Winner, *Autonomous Technology: Technics-out-of-Control as a Theme in Political Thought* (Cambridge, MA: MIT Press, 1977); Leo Marx, "The Idea of 'Technology' and Postmodern Pessimism," in *Does Technology Drive History? The Dilemma of Technological Determinism*, ed. Merritt Roe Smith and Leo Marx (Cambridge, MA: MIT Press, 1994),

237–258; David Landes, *The Unbound Prometheus: Technological Change and Industrial Development in Western Europe from 1750 to the Present* (Cambridge: Cambridge University Press, 1969); David E. Nye, *American Technological Sublime* (Cambridge, MA: MIT Press, 1994).

14 Lorraine J. Daston and Peter Galison, *Objectivity* (New York: Zone Books, 2007).

15 Also see Lorraine J. Daston, "Objectivity and the Escape from Perspective," *Social Studies of Science* 22 (1992): 597–618.

16 Daston and Galison, *Objectivity*, 123.

17 Lorraine Daston, "The Moral Economy of Science," *Osiris* 10 (1995): 3–24.

18 Eric M. Eisenberg, "Ambiguity as Strategy in Organisational Communication," *Communication Monographs* 51, no. 3 (1984): 227–242.

19 Also see Claudia Aradau and Tobias Blanke, "The (Big) Data-Security Assemblage: Knowledge and Critique," *Big Data & Society* 2, no. 2 (2015): 1–12.

20 Catherine Gale, prod. and dir., *The Joy of Data*, aired July 20, 2016, on BBC Four.

21 See Martin Frické, "The Knowledge Pyramid: A Critique of the DIKW Hierarchy," *Journal of Information Science* 35, no. 2 (2009): 131–142.

22 Ernesto Ramirez, "QS Access: Personal Data Freedom," *Quantified Self*, February 11, 2015, http://quantifiedself.com/2015/02/qs-access-personal-data-freedom/.

23 Sara M. Watson, "You Are Your Data," *Slate*, November 12, 2013, www.slate.com/articles/technology/future_tense/2013/11/quantified_self_self_tracking_data_we_need_a_right_to_use_it.html; Gary Wolf, "A Public Infrastructure for Data Access," *Quantified Self*, March 8, 2016, http://quantifiedself.com/2016/03/larry-smarr-interview/.

24 Lisa Gitelman and Virginia Jackson, "Introduction," In *Raw Data Is an Oxymoron*, ed. Lisa Gitelman (Cambridge, MA: MIT Press, 2013), 1–14.

25 The degree and type of human intervention depend on the actual methods involved, of course. It is one thing to manually specify categories in a database and code for automatic sorting of new entries, and it is quite another to train artificial neural nets to develop their own categories as they look for the optimal structure to arrive at desired outcomes (which are, in this case, set manually by human designers). But any specific automated process is part of a larger process in which human coding, selection, categorization, and cleaning of data often plays an important role. Alternatively, could data be "raw" in the sense that it represents the independent objective truth of certain "elementary" natural facts—for example, the weight of an apple. Without delving into the ontological debates, it suffices to point out that the machinic apparatus and its ability to recognize "weight" is, from their inception, designed by humans in terms of what such phenomena mean *for* us.

26 Josh Berson, *Computable Bodies: Instrumented Life and the Human Somatic Niche* (London: Bloomsbury Academic, 2015).

27 For example, Kevin D. Haggerty and Richard V. Ericson, "The Surveillant Assemblage," *British Journal of Sociology* 51, no. 4 (2000): 605–622; John Cheney-Lippold, *We Are Data—Algorithms and the Making of Our Digital Selves* (New

York: New York University Press, 2017); Sara M. Watson, "Data Doppelgängers and the Uncanny Valley of Personalization," *Atlantic*, June 16, 2014, https://www.theatlantic.com/technology/archive/2014/06/data-doppelgangers-and-the-uncanny-valley-of-personalization/372780/; Bernard Harcourt, *Exposed: Desire and Disobedience in the Digital Age* (Cambridge, MA: Harvard University Press, 2015).

28 Daniel Rosenberg, "Data before the Fact," in *Raw Data Is an Oxymoron*, ed. Lisa Gitelman (Cambridge, MA: MIT Press, 2013), 15–40.

29 Respectively, Ed Finn, *What Algorithms Want: Imagination in the Age of Computing* (Cambridge, MA: MIT Press, 2017); N. Katherine Hayles, *My Mother Was A Computer: Digital Subjects and Literary Texts* (Chicago: University of Chicago Press, 1999); David Golumbia, *The Cultural Logic of Computation* (Cambridge, MA: Harvard University Press, 2009).

30 Ludwig Wittgenstein, *On Certainty*, ed. G. E. M Anscombe and G. H von Wright (New York: Harper Torchbooks, 1969), sec. 88.

31 See Schulte, "World-Picture and Mythology," *Inquiry: An Interdisciplinary Journal of Philosophy* 31 (1988): 323–334.

32 Wittgenstein, *On Certainty*, sec. 253.

33 Wittgenstein, *On Certainty*, sec. 103.

34 These regimes, it should be clear, are often maintained not through explicit propositions, as in a code of law or in a textbook on the scientific method; such expressions must themselves be sketched by looking for the shared assumptions and ways of making claims underlying more solid monuments of the regime. It is, Wittgenstein says, a *form of life*. See Wittgenstein, *On Certainty*, sec. 358.

35 This was "Project X" or the "Long Lines Building," a concrete-and-granite skyscraper in Manhattan that reveals nothing about its identity to the city around it. Snowden-leaked files strongly suggest that it was used as an NSA surveillance site for housing and tapping into phone call routing systems. See Ryan Gallagher and Henrik Moltke, "Titanpointe: The NSA's Spy Hub in New York, Hidden in Plain Sight," *The Intercept*, November 16, 2016, https://theintercept.com/2016/11/16/the-nsas-spy-hub-in-new-york-hidden-in-plain-sight/.

36 Ian Hacking, "Statistical Language, Statistical Truth and Statistical Reason: The Self-Authentification of a Style of Scientific Reasoning," in *Social Dimensions of Science*, ed. Ernan McMullin (Notre Dame, IN: University of Notre Dame Press, 1992), 130–157.

37 David Rabouin, "Styles in Mathematical Practice," in *Cultures Without Culturalism: The Making of Scientific Knowledge*, ed. Karine Chemla and Evelyn Fox Keller, 196–223 (Durham, NC: Duke University Press, 2017), 207.

38 A point also made frequently in science studies. See Bruno Latour, *Pandora's Hope: Essays in the Reality of Science Studies* (Cambridge, MA: Harvard University Press, 1999).

39 See, for example, Michael Betancourt, "The Demands of Agnotology::Surveillance," *Ctheory*, 2014, http://ctheory.net/ctheory_wp/the-

demands-of-agnotologysurveillance; Shoshana Zuboff, "Big Other: Surveillance Capitalism and the Prospects of an Information Civilization," *Journal of Information Technology* 30, no. 1 (2015): 75–89.

40 Bernhard Rieder, "Scrutinizing an Algorithmic Technique: The Bayes Classifier as Interested Reading of Reality," *Information, Communication & Society* 20, no. 1 (2017): 100–117.

41 Louise Amoore, "Doubt and the Algorithm: On the Partial Accounts of Machine Learning," *Theory, Culture & Society* 28, no. 6 (2011): 24–43.

42 Orit Halpern, *Beautiful Data: A History of Vision and Reason since 1945* (Durham, NC: Duke University Press, 2014), 26.

43 Jim Thatcher, David O'Sullivan, and Dillon Mahmoudi, "Data Colonialism through Accumulation by Dispossession: New Metaphors for Daily Data," *Environment and Planning D: Society and Space* 34, no. 6 (2016): 991. Also see Nick Srnicek, *Platform Capitalism* (Cambridge: Polity Press, 2017).

44 Mark Andrejevic, "To Preempt a Thief," *International Journal of Communication* 11 (2017): 890.

45 These public–private partnerships are animated not only by the commercial logic but also by the state's interests in securitization. The US government, frequently an open admirer of the Valley during much of this period, repeatedly sought its technologies; In-Q-Tel also funded a mapping company that Google would purchase to build Google Earth. Shoshana Zuboff, *The Age of Surveillance Capitalism: The Fight for a Human Future at the New Frontier of Power* (New York: PublicAffairs, 2019), 117.

46 Peter Waldman, Lizette Chapman, and Jordan Robertson, "Palantir Knows Everything About You," *Bloomberg Businessweek*, April 19, 2018, https://www.bloomberg.com/features/2018-palantir-peter-thiel/.

47 Erika Pearson, "Smart Objects, Quantified Selves, and a Sideways Flow of Data" (presentation at *ICA 2016*, Fukuoka, Japan, 2016).

48 For example, Linda Ackerman, "Mobile Health and Fitness Applications and Information Privacy," 2013 (San Diego, CA: Privacy Rights Clearinghouse); Kate Kaye, "FTC: Fitness Apps Can Help You Shred Calories—and Privacy," *AdAge*, May 7, 2014, http://adage.com/article/privacy-and-regulation/ftc-signals-focus-health-fitness-data-privacy/293080/.

49 For example, Zuboff, *The Age of Surveillance Capitalism*.

50 Zuboff, "Big Other," 75.

51 John Bellamy Foster and Robert W. McChesney, "Surveillance Capitalism: Monopoly-Finance Capital, the Military-Industrial Complex, and the Digital Age," *Monthly Review* 66, no. 3 (2014), https://monthlyreview.org/2014/07/01/surveillance-capitalism/.

52 See David Harvey, *Marx, Capital, and the Madness of Economic Reason* (New York: Oxford University Press, 2018).

53 Srnicek, *Platform Capitalism*.

54 David Harvey, *The New Imperialism* (Oxford: Oxford University Press, 2003). Elsewhere, he terms it a madness, a Hegelian bad infinity: Harvey, *Marx, Capital, and the Madness of Economic Reason*, 172–173. The idea has been taken up more commonly with respect to Big Tech data brokers such as Google and Facebook. For more, see Thatcher, O'Sullivan, and Mahmoudi, "Data Colonialism through Accumulation by Dispossession"; Zuboff, *The Age of Surveillance Capitalism*, 99.

55 For example, see Zygmunt Bauman and David Lyon, *Liquid Surveillance: A Conversation* (Cambridge: Polity, 2013), 51–52.

56 Rosen and Santesso's historical study also emerges with a definition of surveillance not tied to centralizing, coercive power, but "the monitoring of human activities for the purposes of anticipating or influencing future events." David Rosen and Aaron Santesso, *The Watchman in Pieces—Surveillance, Literature, and Liberal Personhood* (New Haven, CT: Yale University Press, 2013), 10.

57 Namely, Ibrahim Diallo, a contractor who was fired by an errant data input that set in motion a set of termination procedures that neither his manager nor departmental director could stop.

58 For more on this subject, see Safiya Umoja Noble, *Algorithms of Oppression: How Search Engines Reinforce Racism* (New York: New York University Press, 2018), 7.

59 Cheney-Lippold, *We Are Data*, 19.

60 An anxiety that has been with us since at least the later nineteenth century. See Rosalind H. Williams, *The Triumph of Human Empire: Verne, Morris, and Stevenson at the End of the World* (Chicago: University of Chicago Press, 2013).

61 See Mark Fisher, *Capitalist Realism: Is There No Alternative?* (Winchester, UK: O Books, 2009).

CHAPTER 2. THE INDEFINITE ARCHIVE

1 *Last Week Tonight with John Oliver*, featuring John Oliver, Tim Carvell, James Taylor, and Jon Thoday, aired April 5, 2015, on HBO.

2 Timothy Morton, *Hyperobjects: Philosophy and Ecology after the End of the World* (Minneapolis: University of Minnesota Press, 2013).

3 The irony being, of course, that Kant's valorization of "public use of reason" was restricted to actions outside each subject's "private," professional activity, in which the overriding virtue was to obey. One suspects Snowden's actions would not have passed this test. See Immanuel Kant, "An Answer to the Question: What Is Enlightenment?" in *What Is Enlightenment? Eighteenth-Century Answers and Twentieth-Century Questions*, ed. James Schmidt, translated by James Schimdt (Berkeley: University of California Press, 1996), 58–77.

4 Kant, *What Is Enlightenment?*, 63. He was, of course, referring to Julien Offray La Mettrie's *L'Homme machine*, then and now widely seen as a culmination of a radical materialism in the Enlightenment. La Mettrie's invocation of machine extended "a general trend . . . towards rationalisation and secularisation" in the

period. See Jonathan I. Israel, *Radical Enlightenment: Philosophy and the Making of Modernity 1650–1750* (Oxford: Oxford University Press, 2001), 6.

5 Alan Rusbridger, "The Snowden Leaks and the Public," *The New York Review of Books*, November 21, 2013, http://www.nybooks.com/articles/2013/11/21/snowden-leaks-and-public/.

6 Cyrus Farivar, "Snowden Distributed Encrypted Copies of NSA Files across the World," *Wired*, January 6, 2013, http://www.wired.co.uk/news/archive/2013-06/26/edward-snowden-nsa-data-copies.

7 Kim Zetter, "Snowden Smuggled Documents from NSA on a Thumb Drive," *Wired*, June 13, 2013, http://www.wired.com/2013/06/snowden-thumb-drive/.

8 Respectively, Michael B. Kelley, "The Guardian's Bombshell Revelation About NSA Domestic Spying Is Only The Tip Of The Iceberg," *Business Insider*, June 6, 2013, https://www.businessinsider.com/the-impact-of-nsa-domestic-spying-2013-6; Ryan Gallagher, "Latest Documents from Snowden Provide Direct Proof of Unlawful Spying on Americans," *Slate*, August 16, 2013, http://www.slate.com/blogs/future_tense/2013/08/16/latest_snowden_documents_prove_proof_of_unlawful_spying_on_americans.html.

9 Antony Loewenstein, "The Ultimate Goal of the NSA Is Total Population Control," *The Guardian*, July 10, 2014, http://www.theguardian.com/commentisfree/2014/jul/11/the-ultimate-goal-of-the-nsa-is-total-population-control.

10 Matthew Cole and Robert Windrem, "How Much Did Snowden Take? At Least Three Times Number Reported," NBC News, August 30, 2013. http://www.nbcnews.com/news/other/how-much-did-snowden-take-least-three-times-number-reported-f8C11038702; Mark Hosenball, "NSA Chief Says Snowden Leaked up to 200,000 Secret Documents," Reuters, November 14, 2013, http://www.reuters.com/article/us-usa-security-nsa-idUSBRE9AD19B20131114; John Borland, "Glenn Greenwald: 'A Lot' More NSA Documents to Come," *Wired*, December 27, 2013, http://www.wired.com/2013/12/greenwald-lot-nsa-documents-come/; Adam Goldman, "The NSA Has No Idea How Much Data Edward Snowden Took Because He Covered His Digital Tracks," *Business Insider*, August 24, 2013, http://www.businessinsider.com/edward-snowden-covered-tracks-2013-8; Steve Almasy, "Journalist: Snowden Has More Documents that Could Harm U.S.," CNN, July 15, 2013. www.cnn.com/2013/07/14/politics/nsa-leak-greenwald/.

11 Glenn Greenwald, "'Explosive' NSA Spying Reports Are Imminent," *Der Spiegel*, July 19, 2013, http://www.wired.com/2013/12/greenwald-lot-nsa-documents-come/.

12 "Interview: Glenn Greenwald," *The Nation*, aired September 13, 2014, on TV3 (New Zealand).

13 Jason Leopold, "Inside Washington's Quest to Bring Down Edward Snowden," *Vice News*, June 4, 2015, https://news.vice.com/article/exclusive-inside-washingtons-quest-to-bring-down-edward-snowden.

14 Michael B. Kelley, "Snowden Has One Very Important and Potentially Devastating Question to Answer," *Business Insider*, March 19, 2014, http://www.businessinsider.com/snowden-and-military-information-2014-3.

15 Senator Susan Collins raised the figure in the 2014 hearing for the Select Commit-
    tee on Intelligence of the US Senate, which is officially open to the public.

16 For example, Tim Shorrock, "US Intelligence Is More Privatized Than Ever
    Before," *Nation*, September 16, 2015, http://www.thenation.com/article/us-
    intelligence-is-more-privatized-than-ever-before/.

17 Frank Pasquale, *The Black Box Society: The Secret Algorithms that Control Money
    and Information* (Cambridge, MA: Harvard University Press, 2015), 13; James
    Bamford, *The Shadow Factory: The Ultra-Secret NSA from 9/11 to the Eavesdrop-
    ping on America* (New York: Doubleday, 2008), 199.

18 Stephen Jay Gould, *The Mismeasure of Man: The Definitive Refutation to the Argu-
    ment of The Bell Curve* (New York: W. W. Norton & Co., 1996), 106.

19 Theodore M. Porter, *Trust in Numbers: The Pursuit of Objectivity in Science and
    Public Life* (Princeton, NJ: Princeton University Press, 1995), 49.

20 Jackson Lears, *Something for Nothing: Luck in America* (New York: Viking, 2003), 82.

21 Sheila Jasanoff, *Science at the Bar: Law, Science, and Technology in America* (Cam-
    bridge: Harvard University Press, 1995), 128.

22 Oliver et al., *Last Week Tonight with John Oliver.*

23 "Majority Views NSA Phone Tracking as Acceptable Anti-Terror Tactic," Pew
    Research Center, June 10, 2013.

24 Barton Gellman and Matt DeLong, "The NSA's Problem? Too Much Data," *The
    Washington Post*, October 14, 2013, http://apps.washingtonpost.com/g/page/
    world/the-nsas-overcollection-problem/517/; Steven Rich and Matt DeLong,
    "NSA Slideshow on 'The TOR Problem,'" *The Washington Post*, October 4,
    2013, http://apps.washingtonpost.com/g/page/world/nsa-slideshow-on-the-tor-
    problem/499/.

25 Jacques Derrida, *Archive Fever: A Freudian Impression* (Chicago: Chicago
    University Press, 1998). For this reading of Derrida, also see Carolyn Steedman,
    Dust: The Archive and Cultural History (New Brunswick, NJ: Rutgers University
    Press, 2002), 3.

26 For an earlier historical episode in the database as a locus of epistemic desire, see
    Rebecca Lemov, *Database of Dreams: The Lost Quest to Catalog Humanity* (New
    Haven, CT: Yale University Press, 2015).

27 This ambiguity or duality is akin to that of boundary objects in the work of Susan
    Leigh Star. For example, Susan Leigh Star and James R. Griesemer, "Institutional
    Ecology, 'Translations' and Boundary Objects: Amateurs and Professionals in
    Berkeley's Museum of Vertebrate Zoology, 1907–39," *Social Studies of Science* 19,
    no. 3 (1989): 387–420.

28 This is in Latour's sense of the term. See Bruno Latour, "Why Has Critique Run
    out of Steam? From Matters of Fact to Matters of Concern," *Critical Inquiry* 30
    (2004): 225–248.

29 Bernard Harcourt, *Exposed: Desire and Disobedience in the Digital Age* (Cam-
    bridge, MA: Harvard University Press, 2015).

30 Pasquale, *The Black Box Society*, 9.

31 John Cheney-Lippold, *We Are Data—Algorithms and the Making of Our Digital Selves* (New York: New York University Press, 2017), 169–173.

32 Walter Kirn, "If You're Not Paranoid, You're Crazy," The *Atlantic*, November 1, 2015, http://www.theatlantic.com/magazine/archive/2015/11/if-youre-not-paranoid-youre-crazy/407833/.

33 For more on the theme of promiscuous networks, see Wendy Hui Kyong Chun, *Updating to Remain the Same: Habitual New Media* (Cambridge, MA: MIT Press, 2015).

34 More precisely, a recent study has shown how large numbers of porn websites send user data to third-party data brokers including Google and Facebook. Elena Maris, Timothy Libert, and Jennifer Henrichsen, "Tracking Sex: The Implications of Widespread Sexual Data Leakage and Tracking on Porn Websites," *New Media & Society* (2020), https://arxiv.org/abs/1907.06520.

35 See Sam Nichols, "Your Phone Is Listening and It's Not Paranoia," *Vice Media*, June 4, 2018, https://www.vice.com/en_au/article/wjbzzy/your-phone-is-listening-and-its-not-paranoia.

36 Antonia Majaca, "Little Daniel Before the Law: Algorithmic Extimacy and the Rise of the Paranoid Apparatus," *E-Flux* 75 (2016), https://www.e-flux.com/journal/75/67140/little-daniel-before-the-law-algorithmic-extimacy-and-the-rise-of-the-paranoid-apparatus/.

37 Michael Taussig, *Defacement—Public Secrecy and the Labour of the Negative* (Stanford, CA: Stanford University Press, 1999).

38 Carl Freedman describes Freud's paranoia in this way, applying the concept to the analysis of science fiction. See Carl Freedman, "Towards a Theory of Paranoia: The Science Fiction of Philip K. Dick," *Science Fiction Studies* 11, no. 1 (1984): 15–24.

39 "If It Weren't for Edward Snowden Conspiracy Theories Would Still Just Be 'Theories,'" Reddit, March 15, 2015, www.reddit.com.

40 Ole Bjerg and Thomas Presskorn-thygesen, "Conspiracy Theory: Truth Claim or Language Game?," *Theory, Culture & Society* 34, no. 1 (2017): 137–159.

41 Bjerg and Presskorn-thygesen, "Conspiracy Theory," 144.

42 Stef Aupers, "'Trust No One': Modernization, Paranoia and Conspiracy Culture," *European Journal of Communication* 27, no. 1 (2012): 22–34.

43 Notably, although Hofstadter also steers away from the clinical meaning of paranoid as a sickness, he is unapologetic about describing the paranoid style as making a "curious leap in imagination" that ultimately departs from the world of actual facts. See Richard Hofstadter, "The Paranoid Style in American Politics," in *The Paranoid Style in American Politics and Other Essays* (New York: Vintage Books, 1967), 37–38.

44 For a historical analysis of this agency panic at the heart of American paranoia, see Timothy Melley, *Empire of Conspiracy: The Culture of Paranoia in Postwar America* (Ithaca, NY: Cornell University Press, 2000), chap. 6.

45 Tom Cohen, "Military Spy Chief: Have to Assume Russia Knows U.S. Secrets," CNN, March 9, 2014, http://www.cnn.com/2014/03/07/politics/snowden-leaks-

russia/; Andrew Kaczynski, "Former NSA Director on Edward Snowden: 'He's Working for Someone,'" *BuzzFeed*, June 3, 2014, http://www.buzzfeed.com/andrewkaczynski/former-nsa-director-on-edward-snowden-hes-working-for-someon#.pf97er37p; Alex Johnson, "Edward Snowden 'Probably' Not a Russian Spy, New NSA Chief Says," NBC News, June 3, 2014, http://www.nbcnews.com/storyline/nsa-snooping/edward-snowden-probably-not-russian-spy-new-nsa-chief-says-n121926; "Edward Snowden: Whistleblower or Double Agent?," Fox News, June 14, 2013, https://www.foxnews.com/politics/edward-snowden-whistleblower-or-double-agent.

46 Walter Pincus, "Questions for Snowden," *The Washington Post*, July 8, 2013, https://www.washingtonpost.com/world/national-security/questions-for-snowden/2013/07/08/d06eeof8-e428-11e2-80eb-3145e2994a55_story.html.

47 Tor Nørretranders, *The User Illusion—Cutting Consciousness Down to Size* (New York: Viking, 1998).

48 Hito Steyerl, "A Sea of Data: Apophenia and Pattern (Mis-)Recognition," *E-Flux* 72 (2016), https://www.e-flux.com/journal/72/60480/a-sea-of-data-apophenia-and-pattern-mis-recognition/.

49 Tom Siebers, *Cold War Criticism and the Politics of Skepticism* (New York: Oxford University Press, 1993).

50 Respectively, Timothy Melley, *The Covert Sphere: Secrecy, Fiction, and the National Security State* (Ithaca, NY: Cornell University Press, 2012); Michael Schudson, *The Rise of the Right to Know—Politics and the Culture of Transparency, 1945–1975* (Cambridge, MA: Harvard University Press, 2015).

51 Maurice Merleau-Ponty, *Phenomenology of Perception*, trans. Donald A Landes (London: Routledge, 2012), 298–99.

52 For example, Lawrence M. Krauss, "Thinking Rationally about Terror," *The New Yorker*, January 2, 2016, http://www.newyorker.com/news/news-desk/thinking-rationally-about-terror?intcid=mod-most-popular.

53 Reuel Marc Gerecht, "The Costs and Benefits of the NSA," *Weekly Standard*, June 24, 2013, http://www.weeklystandard.com/article/costs-and-benefits-nsa/735246.

54 This definition is part of a stable chain of precedents established by Supreme Court decisions, including *Lujan v. Defenders of Wildlife* (1992), *Warth v. Seldin* (1975) and *Sierra Club v. Morton* (1972). For instance, *Jewel v. National Security Agency* (2011), filed by the Electronic Frontier Foundation on the basis of earlier leaks by Mark Klein, was judged upon these terms, and ultimately granted standing on appeal.

55 Similarly, in *ACLU v. Clapper* (2015), the U.S. government sought to have the case thrown out on precisely this lack of standing, and nearly succeeded; the case is still in process at time of this writing.

56 *Spokeo v. Robins* (2016), for instance, stated that the disclosure of previously private information itself does not count as concrete injury and that the abuse of that information, or at least emotional injury as consequence, must be separately demonstrated. The legal threshold for what aspect of data extraction and usage counts as harmful thus remains murky.

57 Ashley Fantz, "NSA Leaker Ignites Global Debate: Hero or Traitor?," CNN, October 6, 2013, http://www.cnn.com/2013/06/10/us/snowden-leaker-reaction/.

58 See Schudson, *The Rise of the Right to Know*.

59 Consider one of Barack Obama's first communications as the president of the United States; see Barak Obama, "Memorandum for the Heads of Executive Departments and Agencies: Transparency and Open Government," The White House, January 21, 2009. https://www.whitehouse.gov/the_press_office/TransparencyandOpenGovernment. Its easy equation of transparency with accountability and informed citizenry reflects the period's enthusiasm for the idea. Also see Mikkel Flyverbom, "Transparency: Mediation and the Management of Visibilities," *International Journal of Communication* 10, no. 1 (2016): 110–122.

60 Claire Birchall, "Introduction to 'Secrecy and Transparency': The Politics of Opacity and Openness," *Theory, Culture & Society* 28, no. 7–8 (2012): 7–25.

61 Bruno Latour, "What If We Talked Politics a Little?," *Contemporary Political Theory* 2, no. 2 (2003): 147.

62 Also see Gianni Vattimo, *The Transparent Society* (Baltimore: Johns Hopkins University Press, 1992).

63 Geoffrey Bennington, "Kant's Open Secret," *Theory, Culture & Society* 28, no. 7–8 (January 2012): 30.

64 Schudson, *The Rise of the Right to Know*, ch. 5.

65 Chun, "Crisis, Crisis, Crisis, or Sovereignty and Networks"; Wendy Hui Kyong Chun, *Updating to Remain the Same: Habitual New Media* (Cambridge, MA: MIT Press, 2015).

66 Eva Horn, "Logics of Political Secrecy," *Theory, Culture & Society* 28, no. 7–8 (2012): 118.

67 Peter Sloterdijk, *Critique of Cynical Reason*, trans. Michael Eldred (Minneapolis: University of Minnesota Press, 1987).

68 As McLuhan famously said of the electric light. Marshall McLuhan, *Understanding Media: The Extensions of Man* (New York: Mentor, 1964), 8.

69 Schudson, *The Rise of the Right to Know*.

70 David Zaret, *Origins of Democratic Culture: Printing, Petitions, and the Public Sphere in Early-Modern England* (Princeton, NJ: Princeton University Press, 2000), 59.

71 Stories of James V's disguised outings, recurring in various forms ever since the years of his reign, were themselves part of a wider folklore motif of the king in disguise. The accuracy of such stories as pertaining to a specific monarch are generally unverifiable, but they, as a whole, gestured to a popular belief that positioned the king as monitoring (benevolently) his subjects. For example, Stevenson, "'The Gudeman of Ballangeich': Rambles in the Afterlife of James V," *Folklore* 115, no. 2 (2004): 198.

72 Frank Bannister and Regina Connolly, "The Trouble with Transparency: A Critical Review of Openness in e-Government," *Policy & Internet* 3, no. 1 (2011): 158–187.

73 Aleecia McDonald and Lorrie Faith Cranor, "The Cost of Reading Privacy Policies," *I/S—A Journal of Law and Policy for the Information Society* 4, no. 3 (2008): 1–22.

74 For a similar translation of externalities onto an epistemological context, see Naomi Oreskes and Erik M. Conway, *Merchants of Doubt: How a Handful of Scientists Obscured the Truth on Issues from Tobacco Smoke to Global Warming* (New York: Bloomsbury Press, 2010), 237.

75 Joshua Reeves, *Citizen Spies: The Long Rise of America's Surveillance Society* (New York: New York University Press, 2017).

76 For example, Ben Tarnoff, "The Data Is Ours!," *Logic* 4 (2018): 91–108.

77 Mary Douglas, "Dealing with Uncertainty," *Ethical Perspectives* 8, no. 3 (2001): 145.

78 Thomas Nagel, "Concealment and Exposure," *Philosophy & Public Affairs* 27, no. 1 (1998): 3–30.

79 John Cassidy, "Why Edward Snowden Is a Hero," *New Yorker*, June 10, 2013, http://www.newyorker.com/news/john-cassidy/why-edward-snowden-is-a-hero; Shami Chakrabarti, "Let Me Be Clear—Edward Snowden Is a Hero," *The Guardian*, June 14, 2015, https://www.theguardian.com/commentisfree/2015/jun/14/edward-snowden-hero-government-scare-tactics.

80 Fred Fleitz, "Snowden Is a Traitor and a Fraud, Period," *National Review*, September 16, 2016, http://www.nationalreview.com/article/440113/edward-snowden-report-house-intelligence-committee-confirms-he-shouldnt-be-pardoned; Matt Williams, "Edward Snowden Is a 'Traitor' and Possible Spy for China—Dick Cheney," *The Guardian*, June 16, 2013, https://www.theguardian.com/world/2013/jun/16/nsa-whistleblower-edward-snowden-traitor-cheney.

81 Nate Fick, "Was Snowden Hero or Traitor? Perhaps a Little of Both," *The Washington Post*, January 19, 2017, https://www.washingtonpost.com/opinions/was-snowden-hero-or-traitor-perhaps-a-little-of-both/2017/01/19/a2b8592e-c6f0-11e6-bf4b-2c064d32a4bf_story.html?utm_term=.28a8e28234ab.

82 Nagel, "Concealment and Exposure."

83 Arthur Conan Doyle, *The Sign of the Four* (Stilwell, KS: Digireads.com, 2005), 36.

CHAPTER 3. RECESSIVE OBJECTS

1 Shane Harris, *The Watchers: The Rise of America's Surveillance State* (New York: Penguin Press, 2010), ch. 16.

2 Peter Maass, "Inside NSA, Officials Privately Criticize 'Collect It All' Surveillance," *The Intercept*, May 28, 2015, https://theintercept.com/2015/05/28/nsa-officials-privately-criticize-collect-it-all-surveillance/.

3 Jorge Luis Borges, "The Library of Babel," in *Jorge Luis Borges: Collected Fictions*, trans. Andrew Hurley (New York: Penguin Books, 1998), 112–118.

4 Popularized via Paul Lazarsfeld and Robert Merton, "Mass Communication, Popular Taste and Organized Social Action," in *The Process and Effects of Mass Communication*, ed. Wilbur Schramm and Donald F. Roberts, rev. ed. (Chicago: University of Illinois Press, 1971), 554–578.

5 Elizbeth E. Joh, "Feeding the Machine: Policing, Crime Data, & Algorithms," *William & Mary Bill of Rights Journal* 26, no. 3 (2017): 287–306.

6 Hans Ulrich Gumbrecht, *Production of Presence: What Meaning Cannot Convey* (Stanford, CA: Stanford University Press, 2003), 67–8.

7 This is the source of phenomenology's intractable difficulty: How can we "know" a process that stays out of our experience and structures it from backstage? The problem of presence, and the problem of knowing and articulating this presence, has been central to the entire project of phenomenology, from Husserl to Merleau-Ponty's "invisible."

8 Brian Rotman, *Becoming Beside Ourselves: The Alphabet, Ghosts, and Distributed Human Being* (Durham, NC: Duke University Press, 2008); Thrift, "Lifeworld Inc—and What to Do about It," *Environment and Planning D: Society and Space* 29, no. 1 (2011): 5–26.

9 See Lars Frers, "The Matter of Absence," *Cultural Geographies* 20, no. 4 (2013): 431–445; Jean-Michel Saury, "The Phenomenology of Negation," *Phenomenology and the Cognitive Sciences* 8, no. 2 (2008): 245–260. These hybrid deployments differentiate our use of presence from the more monolithic conceptualization that Derrida accused Husserl of in the former's famous criticism of the "metaphysics of presence." Also see Wolfgang Walter Fuchs, *Phenomenology and the Metaphysics of Presence* (The Hague: Martinus Nijhoff, 1976).

10 Sianne Ngai, *Ugly Feelings* (Cambridge, MA: Harvard University Press, 2005), 226–228.

11 Here, I draw on Merleau-Ponty's use of *équipement*: a term that freely traverses the typical distinctions between concepts and things, discursive figurations and material ones. With regard to language, he says, "I reach back for the word as my hand reaches towards the part of my body which is being pricked; the word has a certain location in my linguistic world, and is part of my equipment [il fait partie de mon équipement]." See Maurice Merleau-Ponty, *Phenomenology of Perception*, trans. Donald A Landes London: Routledge, 2012), 210.

12 Immanuel Kant, "An Answer to the Question: What Is Enlightenment?" in *What Is Enlightenment? Eighteenth-Century Answers and Twentieth-Century Questions*, ed. James Schmidt, translated by James Schimdt (Berkeley: University of California Press, 1996), 58–77.

13 Mark B. N. Hansen, *Feed-Forward: On the Future of Twenty-First-Century Media* (Chicago: University of Chicago Press, 2015), 37.

14 Hansen, *Feed-Forward*, 87.

15 In particular, N. Katherine Hayles, *Unthought: The Power of the Cognitive Nonconscious* (Chicago: University of Chicago Press, 2017).

16 Siegfried Zielinski, *Deep Time of the Media: Toward an Archaeology of Hearing and Seeing by Technical Means* (Cambridge, MA: MIT Press, 2006), 6.

17 Eleanor Hill, "Joint Inquiry Staff Statement, Hearing on the Intelligence Community's Response to Past Terrorist Attacks Against the United States from February 1993 to September 2001," 2002, http://fas.org/irp/congress/2002_hr/100802hill.html.

18 Whereas state sponsorship is the key explanation for rise and fall of terrorist attacks in the early 1990s, the reports later in the decade would place greater emphasis on radical Islamic groups, "freelance, transnational terrorists," and eventually, "loosely organized, international networks of terrorists." See, respectively, US Department of State, "Patterns of Global Terrorism, 1995," 1996, https://fas.org/irp/threat/terror_95/index.html; US Department of State, "1996 Patterns of Global Terrorism Report," 1997, https://www.fbi.gov/file-repository/stats-services-publications-terror_96.pdf/view; US Department of State, "Patterns of Global Terrorism 1999," 2000, https://www.fbi.gov/stats-services/publications/terror_99.pdf.

19 National Commission on Terrorist Attacks upon the United States, "The 9/11 Commission Report: Final Report of the National Commission on Terrorist Attacks upon the United States: Executive Summary," 2004, http://govinfo.library.unt.edu/911/report/911Report_Exec.pdf.

20 See Timothy Melley, *The Covert Sphere: Secrecy, Fiction, and the National Security State* (Ithaca, NY: Cornell University Press, 2012).

21 John Poindexter, "Information Awareness Office Overview" (presentation at DARPATech 2002, Anaheim, CA).

22 This hunger is not necessarily a default mode for intelligence agencies. Michael Hayden, then director of the NSA, was hesitant to support metadata surveillance tools such as ThinThread before September 11, partly out of a desire to avoid another Watergate (see Bamford, *The Shadow Factory: The Ultra-Secret NSA from 9/11 to the Eavesdropping on America* [New York: Doubleday, 2008]).

23 Ellen Nakashima and Joby Warrick, "For NSA Chief, Terrorist Threat Drives Passion to 'Collect It All,'" *The Washington Post*, July 14, 2013. www.washingtonpost.com; Glenn Greenwald, *No Place to Hide: Edward Snowden, the NSA and the Surveillance State* (London: Penguin Books, 2014), 96.

24 Dan McQuillan, "Algorithmic States of Exception," *European Journal of Cultural Studies* 18, no. 4–5 (2015): 564–576.

25 Grégoire Chamayou has forcefully argued that such predictors have not been shown to exist and that for them to exist, they would have to predict human intentionality on the basis of essentially behavioral data. See Grégoire Chamayou, "Oceanic Enemy: A Brief Philosophical History of the NSA," *Radical Philosophy* 191, no. July (2015): 4; Cheney-Lippold, *We Are Data—Algorithms and the Making of Our Digital Selves* (New York: New York University Press, 2017), 43–45.

26 Meg Stalcup, "Policing Uncertainty: On Suspicious Activity Reporting," in *Modes of Uncertainty—Anthropological Cases*, ed. Limor Samimian-Darash and Paul Rabinow (Chicago: Chicago University Press, 2015), 80.

27 Bruce Schneier, *Data and Goliath—The Hidden Battles to Collect Your Data and Control Your World* (New York: W. W. Norton, 2015), 138.

28 For example, Alice Goffman, *On the Run: Fugitive Life in an American City* (Chicago: University of Chicago Press, 2014).

29 For example, Human Rights Watch and ACLU, *With Liberty to Monitor All: How Large-Scale US Surveillance Is Harming Journalism, Law, and American Democracy*

(New York: Human Rights Watch, 2014). Meanwhile, it has been suggested that now Snowden has brought so much scrutiny to agencies such as the NSA and GCHQ, the organizations are now seeking to use this new visibility to their advantage.

30 This epigraph splices actual comments sourced from experts and acquaintances quoted by the media (Matt Stout and Donna Goodison, "Dzhokhar Tsarnaev Loves Pot, Wrestling Say Friends," *Boston Herald*, April 20, 2013, http://www.bostonherald.com/news_opinion/local_coverage/2013/04/dzhokhar_tsarnaev_loves_pot_wrestling_say_friends; "Dzhokhar and Tamerlan: A Profile of the Tsarnaev Brothers," CBS News, April 23, 2013, https://www.cbsnews.com/news/dzhokhar-and-tamerlan-a-profile-of-the-tsarnaev-brothers/; Sarah Coffey, Patricia Wen, and Matt Carroll, "Bombing Suspect Spent Wednesday as Typical Student," *Boston Globe*, April 19, 2013, https://www.bostonglobe.com/metro/2013/04/19/bombing-suspect-attended-umass-dartmouth-prompting-school-closure-college-friend-shocked-charge-boston-marathon-bomber/8gbczia4qBiWMAP-oSQhViO/story.html; Barney Henderson, "Boston Marathon Bombs: Suspect Captured—April 20 as It Happened," *The Telegraph*, April 20, 2013, http://www.telegraph.co.uk/news/worldnews/northamerica/usa/10007370/Boston-Marathon-bombs-suspect-captured-April-20-as-it-happened.html; Janet Reitman, "Jahar's World," *Rolling Stone*, July 17, 2013, http://www.rollingstone.com/culture/news/jahars-world-20130717), with minor grammatical alterations for consistency. Put together, they depict no clear explanation of Jahar or his brother, Tamerlan, but questions: What does it mean to "know" someone—as a friend, a colleague, a neighbor? And what does it mean to turn such knowledge into data for predicting terrorist intent?

31 Piotr M. Szpunar, "From the Other to the Double: Identity in Conflict and the Boston Marathon Bombing," *Communication, Culture & Critique* 9, no. 4 (2015): 577–594.

32 Quotations from Camilla Schick and Stephen Castle, "'I Trusted Him': London Attacker Was Friendly With Neighbors," *The New York Times*, June 5, 2017, https://www.nytimes.com/2017/06/05/world/europe/london-attack-theresa-may.html?action=click&contentCollection=Europe&module=RelatedCoverage&region=EndOfArticle&pgtype=article.

33 "Two London Attackers Named by Police," BBC News, June 5, 2017, https://www.bbc.com/news/uk-40165646.

34 Michel Foucault, *The Punitive Society: Lectures at the College de France 1972–1973*, ed. Bernard E. Harcourt (New York: Palgrave Macmillan, 2015).

35 Fred Burton and Scott Stewart, "The 'Lone Wolf' Disconnect," *STRATFOR*, January 30, 2008, https://www.stratfor.com/weekly/lone_wolf_disconnect.

36 This is part of a pattern: killers in this period were quickly labeled lone wolves, swelling the ranks of the category, before further information contradicted the designation. See Daveed Gartenstein-Ross and Nathaniel Barr, "The Myth of Lone-Wolf Terrorism," *Foreign Affairs*, July 26, 2016, https://www.foreignaffairs.com/articles/western-europe/2016-07-26/myth-lone-wolf-terrorism.

37 For example, Ramon Spaaij, *Understanding Lone Wolf Terrorism: Global Patterns, Motivations and Prevention* (Dordrecht: Springer, 2012).

38 For more, see "Lone Wolf Tactical Concept," *Aryan Vanguard*, June 2012, http://aryanvanguard.blogspot.com/2012/06/lone-wolf-tactical-concept.html; "Tom Metzger," Anti-Defamation League, accessed February 20, 2016, http://archive.adl.org/learn/ext_us/tom-metzger/ideology.html?LEARN_Cat=Extremism&LEARN_SubCat=Extremism_in_America&xpicked=2&item=7. This development had its own context: far-right figures such as Louis Beam in the 1980s had already been touting "leaderless resistance" as the next step to their struggle.

39 Jeffrey D. Simon, *Lone Wolf Terrorism: Understanding the Growing Threat* (New York: Prometheus Books, 2013).

40 Also see George Michael, *Lone Wolf Terror and the Rise of Leaderless Resistance*, trans. Donald A Landes (London: Routledge, 2012).

41 Gary Fields and Evan Perez, "FBI Seeks to Target Lone Extremists," *Wall Street Journal*, June 15, 2009, http://www.wsj.com/articles/SB124501849215613523.

42 Paul Gill, John Horgan, and Paige Deckert, "Bombing Alone: Tracing the Motivations and Antecedent Behaviors of Lone-Actor Terrorists," *Journal of Forensic Sciences* 59, no. 2 (March 2014): 425–435.

43 Charles A. Eby, "The Nation that Cried Lone Wolf: A Data-Driven Analysis of Individual Terrorists in the United States Since 9/11" (thesis, Naval Postgraduate School, Monterey, CA, 2012), 11.

44 Respectively, Nathan R. Springer, "Patterns of Radicalisation: Identifying the Markers and Warning Signs of Domestic Lone Wolf Terrorists in Our Midst" (thesis, Naval Postgraduate School, Monterey, CA, 2009); Lisa Kaati and Pontus Svenson, "Analysis of Competing Hypothesis for Investigating Lone Wolf Terrorists," in *European Intelligence and Security Informatics Conference* (Washington, DC: IEEE Computer Society, 2011), 295–299.

45 Gill, Horgan, and Deckert, "Bombing Alone."

46 Joel Brynielsson, Andreas Horndahl, Fredrik Johannson, Lisa Kaati, Christian Martenson, and Pontus Svenson, "Analysis of Weak Signals for Detecting Lone Wolf Terrorists," in *Intelligence and Security Informatics Conference*, ed. Nasrullah Memon and Daniel Zeng (Los Alamitos, CA: IEEE Computer Society, 2012), 197–204.

47 Spaaij, *Understanding Lone Wolf Terrorism*.

48 See journalist Robert Evans's analysis: Evans, "Shitposting, Inspirational Terrorism, and the Christchurch Mosque Massacre," *bellingcat*, March 15, 2019, https://www.bellingcat.com/news/rest-of-world/2019/03/15/shitposting-inspirational-terrorism-and-the-christchurch-mosque-massacre/.

49 See Edwin Bakker and Beatrice de Graaf, "Lone Wolves: How to Prevent This Phenomenon?"; Eby, "The Nation that Cried Lone Wolf"; Gill, Horgan, and Deckert, "Bombing Alone"; Spaaij, *Understanding Lone Wolf Terrorism*.

50 For example, Clark McCauley and Sophia Moskalenko, "Toward a Profile of Lone Wolf Terrorists: What Moves an Individual From Radical Opinion to Radical Ac-

tion," *Terrorism and Political Violence* 26, no. 1 (2014): 69–85; Raffaello Pantucci, *A Typology of Lone Wolves: Preliminary Analysis of Lone Islamist Terrorists* (London: The International Centre for the Study of Radicalisation and Political Violence, 2011), http://mediafieldsjournal.squarespace.com/rise-of-the-imsi-catcher/.

51 For example, J. Oliver Conroy, "They Hate the US Government, and They're Multiplying: The Terrifying Rise of 'Sovereign Citizens,'" *The Guardian*, May 15, 2017, https://www.theguardian.com/world/2017/may/15/sovereign-citizens-rightwing-terrorism-hate-us-government.

52 Kaati and Svenson, "Analysis of Competing Hypothesis."

53 Cora Currier, "48 Questions the FBI Uses to Determine If Someone Is a Likely Terrorist," *The Intercept*, February 13, 2017, https://theintercept.com/2017/02/13/48-questions-the-fbi-uses-to-determine-if-someone-is-a-likely-terrorist/.

54 This is not, of course, unusual in the business. For instance, American parole systems often turn to LSI-R, a multifactor system for predicting recidivism that compiles data from a whole array of life impressions—including frequent address changes and social isolation. Bernard Harcourt, *Against Prediction—Profiling, Policing, and Punishing in an Actuarial Age* (Chicago: Chicago University Press, 2007), 81.

55 Amy Davidson, "The N.S.A.'s Spying on Muslim-Americans," *The New Yorker*, July 10, 2014, http://www.newyorker.com/news/amy-davidson/the-n-s-a-s-spying-on-muslim-americans.

56 "Rolling Stone Defends Boston Bomb Suspect Cover," BBC News, July 17, 2013, https://www.bbc.com/news/world-us-canada-23351317.

57 Jasbin K. Puar and Amit S. Rai, "Monster, Terrorist, Fag: The War on Terrorism and the Production of Docile Patriots," *Social Text* 20, no. 3 (2002): 117–148.

58 For example, Mark Andrejevic, "To Preempt a Thief," *International Journal of Communication* 11 (2017): 879–896.

59 Judith Butler, *The Psychic Life of Power* (Stanford, CA: Stanford University Press, 1997).

60 Sarah Brayne describes this as an unequal broadening of the dragnet. See Sarah Brayne, "Big Data Surveillance: The Case of Policing," *American Sociological Review* 82, no. 5 (2017): 998.

61 Daesh itself soon claimed Paddock as its "soldier," but the FBI was equally quick to dismiss the connection. Given Daesh's historical propensity toward overcrediting itself, its claims on Paddock appear more opportunistic than genuine.

62 Moustafa Bayoumi, "What's a 'Lone Wolf'? It's the Special Name We Give White Terrorists," *The Guardian*, October 2, 2017, https://www.theguardian.com/commentisfree/2017/oct/04/lone-wolf-white-terrorist-las-vegas; Shaun King, "The White Privilege of the 'Lone Wolf' Shooter," *The Intercept*, October 2, 2017, https://theintercept.com/2017/10/02/lone-wolf-white-privlege-las-vegas-stephen-paddock/; Khaled Beydoun, "'Lone Wolf': Our Stunning Double Standard When It Comes to Race and Religion," *The Washington Post*, October 2, 2017, https://www.washingtonpost.com/news/acts-of-faith/wp/2017/10/02/lone-wolf-

our-stunning-double-standard-when-it-comes-to-race-and-religion/?utm_
term=.1e17de8of30e.

63 "Las Vegas Shootings: Is the Gunman a Terrorist?," BBC News, October 3, 2017,
https://www.bbc.com/news/world-us-canada-41483943.

64 Gilles Deleuze, "Postscript on the Societies of Control," *October* 59 (1992): 3–7.

CHAPTER 4. DATA'S INTIMACY

1 Epigraph from Nick Wingfield, "Gauging the Natural, and Digital, Rhythms
of Life," *The New York Times*, June 19, 2013, http://bits.blogs.nytimes.
com/2013/06/19/gauging-the-natural-and-digital-rhythms-of-life/?_r=0.

2 Lasse Leppakorpi, "Beddit Presentation" (presentation given at MoneyTalks, Tampere, FL, February 10, 2011, uploaded to SlideShare, February 14, 2011), https://
www.slideshare.net/TechnopolisOnline/beddit-presentation.

3 Ariel Garten and Mikey Siegel, "Brainwave Technology and Consciousness Hacking"
(presentation at Quantified Self 2015 Conference, San Francisco, June 18–20, 2015).

4 Paraphrased from Brian Massumi, *Parables for the Virtual—Movement, Affect,
Sensation* (Durham, NC: Duke University Press, 2002), 13.

5 Dawn Nafus and Jamie Sherman, "This One Does Not Go Up to 11: The Quantified Self Movement as an Alternative Big Data Practice," *International Journal of
Communication* 8 (2014): 1784–1794. Also see Sara M. Watson, "Living with Data:
Personal Data Uses of the Quantified Self" (master's thesis, University of Oxford,
2013); Deborah Lupton, *The Quantified Self: A Sociology of Self-Tracking* (Cambridge: Polity Press, 2016).

6 Emblematized by lifelogging fantasies of, for instance, recording and passing on
important life moments for one's future self or even one's children. For more, see
Rebecca Lemov, "Archives-of-Self: The Vicissitudes of Time and Self in a Technologically Determinist Future," in *Science in the Archives: Pasts, Presents, Futures*,
ed. Lorraine Daston (Chicago: University of Chicago Press, 2017), 247–270.

7 Paul Erickson, Judy L. Klein, Lorraine Daston, Rebecca Lemov, Thomas Sturm,
and Michael D. Gordin, *How Reason Almost Lost Its Mind* (Chicago: University of
Chicago Press, 2013), 8.

8 Carlo Ginzburg, "Morelli, Freud and Sherlock Holmes: Clues and Scientific
Method," *History Workshop Journal* 9 (1980): 11.

9 For a more detailed critique along these lines, see Adam Mackenzie, "The Production of Prediction: What Does Machine Learning Want?," *European Journal of
Cultural Studies* 18, no. 4–5 (2015): 429–445.

10 Kevin Kelly, "What Is the Quantified Self?," *Quantified Self*, October 5, 2007,
https://web.archive.org/web/20150408202734/http://quantifiedself.com/2007/10/
what-is-the-quantifiable-self/.

11 Also see Lupton, *The Quantified Self*.

12 "Wearables Market to Be Worth $25 Billion by 2019," CCS Insight, September 1,
2015, https://www.ccsinsight.com/press/company-news/2332-wearables-market-
to-be-worth-25-billion-by-2019-reveals-ccs-insight/.

13 "Worldwide Wearables Market Soars in the Third Quarter as Chinese Vendors Challenge the Market Leaders, According to IDC,"a; Lupton, "Quantifying the Body: Monitoring and Measuring Health in the Age of MHealth Technologies," *Critical Public Health* 23, no. 4 (2013): 393–403.

14 Although the technologies discussed in this book are further subdivided, the family resemblances across multiple types of sensors and trackers at the level of usage is also well documented; for example, see Mark Andrejevic and Mark Burdon, "Defining the Sensor Society," *Television & New Media* 16, no. 1 (2015): 19–36.

15 Also see Susan Elizabeth Ryan, *Garments of Paradise: Wearable Discourse in the Digital Age* (Cambridge, MA: MIT Press, 2014).

16 "Smart sensors," typically include both sensors for input, such as light levels or acceleration, and microprocessors for handling signal input and algorithmic analysis before the data leaves the unit had achieved the low power requirements, small size, and affordability conducive to widespread deployment.

17 One example of the Internet of Things before the internet as we know it is *domotique*, a French neologism of *domus* + *informatique* where devices such as computerized thermostats were networked to the nation's Minitel system. See Julien Mailland and Kevin Driscoll, *Minitel: Welcome to the Internet* (Cambridge, MA: MIT Press, 2017), 122.

18 Respectively, Frog, "15 Tech Trends that Will Define 2014, Selected By Frog," *Fast Company*, January 8, 2014, http://www.fastcodesign.com/3024464/15-tech-trends-that-will-define-2014-selected-by-frog; Vivek Wadhwa, "Five Innovation Predictions for 2013," *The Washington Post*, January 4, 2013, https://www.washingtonpost.com/national/on-innovations/five-innovation-predictions-for-2013/2013/01/04/f4718be6-55c5-11e2-bf3e-76c0a789346f_story.html; John Brandon, "6 Tech Trends of the Far Future," *Inc.*, May 16, 2013, http://www.inc.com/john-brandon/6-tech-trends-of-the-far-future.html; "The Decades that Invented the Future, Part 12: The Present and Beyond," *Wired*, February 8, 2013, https://www.wired.com/2013/02/the-decades-that-invented-the-future-part-12-the-present-and-beyond/.

19 Mercedes Bunz and Graham Meikle, *The Internet of Things* (Cambridge: Polity Press, 2018).

20 Frank W. Geels and Wim A. Smit, "Lessons from Failed Technological Futures: Potholes in the Road to the Future," in *Contested Futures: A Sociology of Prospective Techno-Science*, ed. Nik Brown, Brian Rappert, and Andrew Webster (Aldershot, UK: Ashgate, 2000), 129–155.

21 Fred Turner, *From Counterculture to Cyberculture: Stewart Brand, the Whole Earth Network, and the Rise of Digital Utopianism* (Chicago: University of Chicago Press, 2006).

22 Natasha Dow Schüll, "Self-Tracking Technology from Compass to Thermostat" (presentation at Streams of Consciousness conference, University of Warwick, Warwick, UK, April 21–22, 2016); Whitney Erin Boesel, "Return of the Quantre-

preneurs," *Society Pages,* Cyborgology section, September 26, 2013, https://theso-cietypages.org/cyborgology/2013/09/26/return-of-the-quantrepreneurs/.

23 Michel Foucault, *The Courage of Truth: Lectures at the Collège de France, 1983–1984,* ed. Frédéric Gros, trans. Graham Burchell (Basingstoke, UK: Palgrave Macmillan, 2011), 3.

24 This is what Foucault calls a regime of truth, "that which constrains individuals to these truth acts, that which defines, determines the form of these acts and establishes their conditions of effectuation and specific effects." Michel Foucault, *On the Government of the Living: Lectures at the Collège de France, 1979–1980* (Basingstoke, UK: Palgrave Macmillan, 2014), 93.

25 For example, Alex Lambert, "Bodies, Mood and Excess," *Digital Culture & Society* 2, no. 1 (2016): 71–88.

26 For example, Michel Foucault, *Security, Territory, Population: Lectures at the Collège de France 1977–1978,* ed. Michel Senellart, trans. Graham Burchell (New York: Palgrave Macmillan, 2004); Evelyn Ruppert, "Population Objects: Interpassive Subjects," *Sociology* 45, no. 2 (2011): 218–233.

27 For example, Laila Zemrani, "Using Self Tracking to Exercise More Efficiently," (presentation at New York Quantified Self Meetup, New York, 2015).

28 Melanie Swan, "The Quantified Self: Fundamental Disruption in Big Data Science and Biological Discovery," *Big Data* 1, no. 2 (2013): 85–99; Gary Wolf, "Tim Ferriss Wants to Hack Your Body," *Wired,* November 29, 2010. http://www.wired.com/2010/11/mf_qa_ferriss/.

29 For example, Sarita Bhatt, "We're All Narcissists Now, and that's a Good Thing," *Fast Company,* September 27, 2013, http://www.fastcoexist.com/3018382/were-all-narcissists-now-and-thats-a-good-thing; Monica Hesse, "Bytes of Life," *The Washington Post,* September 9, 2008, www.wired.com/2010/11/mf_qa_ferriss/; Alicia Morga, "Do You Measure Up?," *Fast Company,* April 5, 2011, http://www.fastcompany.com/1744571/do-you-measure.

30 Ian Hacking, "Statistical Language, Statistical Truth and Statistical Reason: The Self-Authentification of a Style of Scientific Reasoning," in *Social Dimensions of Science,* ed. Ernan McMullin (Notre Dame, IN: University of Notre Dame Press, 1992), 148.

31 Also see Dana Greenfield, "Deep Data: Notes on the n of 1," in *Quantified—Biosensing Technologies in Everyday Life,* ed. Dawn Nafus (Cambridge, MA: MIT Press, 2016), 123–146.

32 Ryan Calo, "Digital Market Manipulation," *George Washington Legal Review* 82, no. 4 (2014): 1010–1011.

33 Also see David Lyon, "Surveillance, Power and Everyday Life," in *Oxford Handbook of Information and Communication Technologies,* ed. Chrisanthi Avgerou, Robin Mansell, Danny Quah, and Roger Silverstone (Oxford: Oxford University Press, 2007), 449–472.

34 Michel Foucault, *Wrong-Doing, Truth-Telling: The Function of Avowal in Justice,* ed. Fabienne Brion and Bernard E. Harcourt (Chicago: University of Chicago Press, 2014), 11–12.

35 Michel Foucault, *Speech Begins after Death* (Minneapolis: University of Minnesota Press, 2017), 35.

36 For example, Daniela Hernandez, "Big Data Is Transforming Healthcare," *Wired*, October 16, 2012, http://www.wired.com/2012/10/big-data-is-transforming-healthcare/; Jill Walker Rettberg, *Seeing Ourselves Through Technology: How We Use Selfies, Blogs and Wearable Devices to See and Shape Ourselves* (Basingstoke, UK: Palgrave Macmillan, 2014).

37 Mark Weiser, "The Computer for the 21st Century," *Scientific American* (September 1991): 94–104.

38 Erich Hörl, "The Technological Condition," *Parrhesia* 22 (2015): 8.

39 See James Ash, *Phase Media: Space, Time and the Politics of Smart Objects* (New York: Bloomsbury Academic, 2018), 5.

40 See Nigel Thrift, "Lifeworld Inc—and What to Do about It," *Environment and Planning D: Society and Space* 29, no. 1 (2011): 5–26; David M. Berry, *The Philosophy of Software: Code and Mediation in the Digital Age* (Basingstoke, UK: Palgrave Macmillan, 2011); Judy Wajcman and Emily Rose, "Constant Connectivity: Rethinking Interruptions at Work," *Organization Studies* 32, no. 7 (2011): 941–961.

41 Ariel Garten in discussion with the author, June 2015.

42 See Megan Garber, "The Ennui of the Fitbit," "The Ennui of the Fitbit." The *Atlantic*, July 10, 2015, https://www.theatlantic.com/technology/archive/2015/07/the-ennui-of-the-fitbit/398129/. (The article was alternatively titled: "Fitbit? More Like Quitbit.")

43 Relatedly, Natasha Dow Schüll identifies what she calls the paradigm of 'data for life': the wearables industry's effort to cultivate a long-term habit of use unanchored from any specific problem or tool. See Natasha Dow Schüll, "Data for Life: Wearable Technology and the Design of Self-Care," *Biosocieties* 11, no. 3 (2016): 317–333.

44 Watson, "Living with Data," 11.

45 Gary Wolf, "Know Thyself: Tracking Every Facet of Life, from Sleep to Mood to Pain, 24/7/365," *Wired*, June 22, 2009, http://www.wired.com/2009/06/lbnp-knowthyself/. For more on this "origin myth," see Minna Ruckenstein and Mika Pantzar, "Beyond the Quantified Self: Thematic Exploration of a Dataistic Paradigm," *New Media & Society* 19, no. 3(2017): 401-418.

46 Gary Wolf, "What Is the Quantified Self?," *Quantified Self*, March 3, 2011, http://quantifiedself.com/2011/03/what-is-the-quantified-self/.

47 Rob Walker, "Wasted Data," *The New York Times*, December 3, 2010, http://www.nytimes.com/2010/12/05/magazine/05FOB-Consumed-t.html?mtrref=undefined&gwh=224196DF50B6FEC7F871890CABC54F5A&gwt=pay.

48 Sam de Brouwer, Linda Avey, Jessica Richman, and Tan Le, "Frontiers of Tracking Health" (presentation at Quantified Self 2015 Conference, San Francisco, June 18–20, 2015).

49 Jacoba Urist, "From Paint to Pixels," *The Atlantic*, May 14, 2015, http://www.theatlantic.com/entertainment/archive/2015/05/the-rise-of-the-data-artist/392399/.

50 For example, José van Dijck, *The Culture of Connectivity: A Critical History of Social Media* (Cambridge: Oxford University Press, 2013); Michael Zimmer, "The Externalities of Search 2.0: The Emerging Privacy Threats When the Drive for the Perfect Search Engine Meets Web 2.0," *First Monday* 13, no. 3 (2008), http://www.theguardian.com/commentisfree/2014/jul/14/fitness-tracker-vagina-quantified-life.

51 See Lucas Mearian, "Insurance Company Now Offers Discounts—If You Let It Track Your Fitbit," *Computerworld*, April 17, 2015, http://www.computerworld.com/article/2911594/insurance-company-now-offers-discounts-if-you-let-it-track-your-fitbit.html; Erika Pearson, "Smart Objects, Quantified Selves, and a Sideways Flow of Data" (presentation at ICA 2016, Fukuoka, Japan, June 9–13, 2016).

52 Hesse, "Bytes of Life."

53 Richard H. Thaler and Cass R. Sunstein, *Nudge: Improving Decisions about Health, Wealth and Happiness* (New York: Penguin Books, 2009).

54 This basic pattern also characterizes the influence of neuroscience and the behavioral sciences on the humanities and other avenues for the analysis of sociocultural phenomena, including the "turn to affect." See Constantina Papoulias and Felicity Callard, "Biology's Gift: Interrogating the Turn to Affect," *Body & Society* 16, no. 1 (2010): 29–56; Ruth Leys, "The Turn to Affect: A Critique," *Critical Inquiry* 37, no. 3 (2011): 434–472; Ben Anderson, "Affect and Biopower: Towards a Politics of Life," *Transactions of the Institute of British Geographers* 37, no. 1 (2012): 30.

55 For another take on this aspect, see John Cheney-Lippold's discussion of the "pro-tocategorical": John Cheney-Lippold, *We Are Data—Algorithms and the Making of Our Digital Selves* (New York: New York University Press, 2017).

56 Steven Jonas, "What We Are Reading," *Quantified Self*, October 26, 2015, http://quantifiedself.com/2015/10/reading-73/.

57 Also see Schüll, "Data for Life."

58 For example, Tarleton Gillespie, "The Politics of 'Platforms,'" *New Media & Society* 12, no. 3 (2010): 347–364.

59 This impression is latent in technology's foundational equivalence with unending and universal progress (e.g., Leo Marx, "Technology: The Emergence of a Hazardous Concept," *Technology and Culture* 51, no. 3 [2010]: 561–577).

60 Also see Donald Mackenzie, "Marx and the Machine," *Technology and Culture* 25, no. 3 (1984): 473–502.

61 For example, Louise Amoore, "Data Derivatives: On the Emergence of a Security Risk Calculus for Our Times," *Theory, Culture & Society* 28, no. 6 (2011): 24–43; Taina Bucher, "Want to Be on the Top? Algorithmic Power and the Threat of Invisibility on Facebook," *New Media & Society* 14, no. 7 (2012): 1164–1180; Tarleton Gillespie, "#Trendingistrending: When Algorithms Become Culture," in *Algorithmic Cultures: Essays on Meaning, Performance and New Technologies*, ed. Robert Seyfert and Jonathan Roberge (Abingdon, UK: Routledge, 2016), 52–75; Frank Pasquale, *The Black Box Society: The Secret Algorithms that Control Money and Information* (Cambridge, MA: Harvard University Press, 2015).

62 Adapted from N. Katherine Hayles, *Unthought: The Power of the Cognitive Non-conscious* (Chicago: University of Chicago Press, 2017), 26.

63 Benjamin Franklin, *The Autobiography of Benjamin Franklin* (London: George Bell & Sons, 1884).

64 Tahl Milburn, "How My Life Automation System Quantifies My Life" (presentation at Quantified Self 2015 Conference, San Francisco, June 18–20, 2015).

65 Ana Viseu and Lucy Suchman, "Wearable Augmentations," in *Technologized Images, Technologized Bodies: Anthropological Approaches to a New Politics of Vision*, ed. Jeanette Edwards, Penelope Harvey, and Peter Wade (New York: Berghahn Books, 2010), 161–84.

66 Sarah Kessler, "Can the Quantified Self Go Too Far?," *Fast Company*, August 19, 2013, www.fastcompany.com.

67 Mark Coeckelbergh, *New Romantic Cyborgs: Romanticism, Information Technology, and the End of the Machine* (Cambridge, MA: MIT Press, 2017).

68 Theodore M. Porter, *Trust in Numbers: The Pursuit of Objectivity in Science and Public Life* (Princeton, NJ: Princeton University Press, 1995).

69 Anna North, "Why You Want an App to Measure Calories but Not Character," *The Washington Post*, August 26, 2014, http://op-talk.blogs.nytimes.com; Maciej Ceglowski, "Haunted by Data" (presentation at the Strata+Hadoop World Conference, New York, September 29–October 1, 2015).

70 For example, Lane Wallace, "The Illusion of Control," *The Atlantic*, May 26, 2010, http://www.theatlantic.com/technology/archive/2010/05/the-illusion-of-control/57294/; Thomas Goetz, "The Diabetic's Paradox," *The Atlantic*, April 1, 2013, http://www.theatlantic.com/health/archive/2013/04/the-diabetics-paradox/274507/.

71 A similar argument was being raised against the use of big data (e.g., danah boyd and Kate Crawford, "Critical Questions for Big Data," *Information, Communication & Society* 15, no. 5 [2012]: 662–679) and increasingly complex models in other knowledge spheres, including finance. See Donald Mackenzie, "Unlocking the Language of Structured Securities," *Financial Times*, August 18, 2010, http://www.ft.com/cms/s/0/8127989a-aae3-11df-9e6b-00144feabdc0.html#axzz3w2fTTGpL.

72 Evgeny Morozov, *To Save Everything, Click Here: The Folly of Technological Solutionism* (New York: Public Affairs, 2013).

73 Nafus would later translate this discussion into a co-authored monograph on self-tracking, warning that self-tracking can "trap" users into a normative mode of self-interrogation. Gina Neff and Dawn Nafus, *Self-Tracking* (Cambridge, MA: MIT Press, 2016), 38–43.

74 Gary Wolf, interview, 8 August 2015.

75 Also see the analysis of exposure in Bernard Harcourt, *Exposed: Desire and Disobedience in the Digital Age* (Cambridge, MA: Harvard University Press, 2015).

76 Jess Zimmerman, "There's a Fitness Tracker for Vaginas. Quantifying Your Life Has Gone Too Far," *The Guardian*, July 14, 2014, http://www.theguardian.com/commentisfree/2014/jul/14/fitness-tracker-vagina-quantified-life.

77 "Pplkpr," *pplkpr.com*, accessed March 31, 2016, www.pplkpr.com.

78 Liz Stinson, "Having a Hard Time Being a Human? This App Manages Friendships for You," *Wired*, January 26, 2015, http://www.wired.com/2015/01/hard-time-human-app-manages-friendships/.

79 Caitlin Dewey, "Everyone You Know Will Be Able to Rate You on the Terrifying 'Yelp for People'—Whether You Want Them to or Not," *The Washington Post*, September 30, 2015, https://www.washingtonpost.com/news/the-intersect/wp/2015/09/30/everyone-you-know-will-be-able-to-rate-you-on-the-terrifying-yelp-for-people-whether-you-want-them-to-or-not/.

80 Tomas Chamorro-Premuzic, "Reputation and the Rise of the 'Rating' Society," *The Guardian*, October 26, 2015, http://www.theguardian.com/media-network/2015/oct/26/reputation-rating-society-uber-airbnb; "Peeple App for Rating Human Beings Causes Uproar," BBC News, October 1, 2015, https://www.bbc.com/news/technology-34415382.

81 Also see Rory Carroll, "Inspector Gadget: How Smart Devices Are Outsmarting Criminals," *The Guardian*, June 23, 2017, https://www.theguardian.com/technology/2017/jun/23/smart-devices-solve-crime-murder-internet-of-things.

82 Parmy Olson, "Fitbit Data Now Being Used in the Courtroom," *Forbes*, November 16, 2014, http://www.forbes.com/sites/parmyolson/2014/11/16/fitbit-data-courtroom-personal-injury-claim.

83 For example, Matthew Jordan and Nikki Pfarr, "Forget the Quantified Self. We Need to Build the Quantified Us," *Wired*, April 4, 2014, http://www.wired.com/2014/04/forget-the-quantified-self-we-need-to-build-the-quantified-us/.

84 Also see Pearson, "Smart Objects."

85 Cathy O'Neill, *Weapons of Math Destruction: How Big Data Increases Inequality and Threatens Democracy* (New York: Crown, 2016), 168.

86 "John Hancock Adds Fitness Tracking to All Policies," BBC News, September 20, 2018, https://www.bbc.com/news/technology-45590293.

87 A historical analysis can be found in Phoebe V. Moore, *The Quantified Self in Precarity—Work, Technology and What Counts* (Abingdon, UK: Routledge, 2018); Martin Upchurch and Phoebe V. Moore, "Deep Automation and the World of Work," in *Humans and Machines at Work: Monitoring, Surveillance and Automation in Contemporary Capitalism*, ed. Phoebe V. Moore, Martin Upchurch, and Xanthe Whittaker (London: Palgrave Macmillan, 2018), 45–71.

88 Ivan Manokha, "New Means of Workplace Surveillance: From the Gaze of the Supervisor to the Digitalization of Employees," *Monthly Review*, February 1, 2019, https://monthlyreview.org/2019/02/01/new-means-of-workplace-surveillance/.

89 Solon Barocas and Karen Levy, "Refractive Surveillance: Monitoring Customers to Manage Workers," *International Journal of Communication* 12 (2007): 1166–1188.

90 Clover, "Clover Provider Manual," 2017.

91 Harcourt, *Exposed*, 199.

92 Siva Vaidhyanathan, *The Googlization of Everything: (And Why We Should Worry)* (Los Angeles: University of California Press, 2011), 113–114.

93 van Dijck, *The Culture of Connectivity*.

94 Andre Oboler, Kristopher Welsh, and Lito Cruz, "The Danger of Big Data: Social Media as Computational Social Science," *First Monday* 17, no. 7 (2012), https://firstmonday.org/article/view/3993/3269/.

95 For surveillance as social sorting, see Zygmunt Bauman and David Lyon, *Liquid Surveillance: A Conversation* (Cambridge: Polity, 2013); Gandy, *The Panoptic Sort: A Political Economy of Personal Information* (Boulder, CO: Westview Press, 1993).

96 For example, Samantha Murphy Kelly, "The Most Connected Man Is You, Just a Few Years From Now," *Mashable*, January 8, 2014, http://mashable.com/2014/08/21/most-connected-man/#I6oSjAremkqw; Sarah Griffiths, "Is This the Most Connected Human on the Planet? Man Is Wired up to 700 Sensors to Capture Every Single Detail of His Existence," *Daily Mail*, March 25, 2014, http://www.dailymail.co.uk/sciencetech/article-2588779/Is-connected-man-planet-Man-wired-700-devices-capture-single-existence.html; Oliver Wainwright, "Rise of the 'Inner-Net': Meet the Most Connected Man on the Planet," *The Guardian*, March 19, 2014, https://www.theguardian.com/artanddesign/architecture-design-blog/2014/mar/19/inner-net-most-connected-man-earth-fitness-trackers-data.

97 For a deeper analysis of this dynamic in the case of productivity and time management techniques, see Melissa Gregg, *Counterproductive: Time Management in the Knowledge Economy* (Durham, NC: Duke University Press, 2018).

98 Chris Dancy, in Moore, *The Quantified Self in Precarity*, 191.

99 Eva Illouz, *Saving the Modern Soul: Therapy, Emotions, and the Culture of Self-Help*, vol. 38 (Berkeley: University of California Press, 2008); Eva Illouz, *Cold Intimacies: The Making of Emotional Capitalism* (Malden, MA: Polity Press, 2007).

100 Another major area for monitoring and rationalizing the human worker during the century was, of course, human relations and the art of management. For studies in this area, see Moore, *The Quantified Self in Precarity*; Rebecca Lemov, "Hawthorne's Renewal: Quantified Total Self," in *Humans and Machines at Work: Monitoring, Surveillance and Automation in Contemporary Capitalism*, ed. Phoebe V. Moore, Martin Upchurch, and Xanthe Whittaker (London: Palgrave Macmillan, 2018), 181–202.

101 Anne Harrington, *The Cure Within: A History of Mind-Body Medicine* (Cambridge, MA: Harvard University Press, 2015).

102 Also see Ruckenstein and Pantzar, "Beyond the Quantified Self."

103 Similar relationships between lifehacking and self-help are detailed in Joseph Reagle, *Hacking Life: Systematized Living and Its Discontents* (Cambridge, MA: MIT Press, 2019).

104 John van Doorn, "An Intimidating New Class: The Physical Elite," *The New York Magazine*, May 29, 1978, http://nymag.com/news/features/49241/.

105 Although data is scarce, one study of self-tracking videos promoted by the QS website showed that they were overwhelmingly male (79%) tech industry professionals (90%). Eun Kyoung Choe, Nicole B. Lee, Bongshin Lee, Wanda Pratt, and Julie A Kientz, "Understanding Quantified-Selfers' Practices in Collecting and

Exploring Personal Data," in *Proceedings of the 32nd Annual ACM Conference on Human Factors in Computing Systems*, ed. Matt Jones, Philippe Palanque, Albrecht Schmidt, and Tovi Grossman (Toronto: ACM, 2014), 1143–1152. Also see Lupton, *The Quantified Self*, 32.

106 Klint Finley, "Interview: Sensor Hacking For Mindfulness with Nancy Dougherty on the New Mindful Cyborgs," *Technoccult*, June 10, 2013, http://technoccult.net/archives/2013/06/10/interview-sensor-hacking-for-mindfulness-with-nancy-dougherty-on-the-new-mindful-cyborgs/.

107 For this specific commentary, see Sarah Banet-Weiser, *Authentic: The Politics of Ambivalence in Brand Culture* (New York: New York University Press, 2012), 189–190.

108 Wolf, "Know Thyself."

109 Mariana Valverde, *Diseases of the Will: Alcohol and the Dilemmas of Freedom* (Cambridge: Cambridge University Press, 1998).

110 For example, Evan Selinger, "Why It's OK to Let Apps Make You a Better Person," *The Atlantic*, March 9, 2012, https://www.theatlantic.com/technology/archive/2012/03/why-its-ok-to-let-apps-make-you-a-better-person/254246/.

111 Most clearly in Michel Foucault, *The Punitive Society: Lectures at the College de France 1972–1973*, ed. Bernard E. Harcourt (New York: Palgrave Macmillan, 2015).

112 For more, see Schüll, "Data for Life."

113 Vinod Khosla, "The Algorithms Are Coming. What's at Stake?"

114 See Bruce Sterling, "The Internet of Things: Quantified Self, IoT, Smart Cities, Smart Cars, Smart Clothes," *Wired*, April 3, 2013, http://www.wired.com/2013/04/the-internet-of-things-quantified-self-iot-smart-cities-smart-cars-smart-clothes/; John C. Havens, *Hacking H(App)Iness: Why Your Personal Data Counts and How Tracking It Can Change the World* (New York: Jeremy P. Tarcher/Penguin, 2014), 58–59.

115 Holman W. Jenkins Jr., "Google and the Search for the Future," *Wall Street Journal*, August 14, 2010, http://www.wsj.com/articles/SB10001424052748704901104575423294099527212.

116 Josh Berson, *Computable Bodies: Instrumented Life and the Human Somatic Niche* (London: Bloomsbury Academic, 2015).

117 Lambert, "Bodies, Mood and Excess."

118 Gregg, *Work's Intimacy*.

119 Also see Moore, *The Quantified Self in Precarity*.

120 Hence, for the subject to be "passionately attached," to become socially intelligible, is not to occupy a static position but constantly orient discourse, affect, behavior, and thereby occupy a trajectory. Also see Butler, *The Psychic Life of Power* (Stanford, CA: Stanford University Press, 1997).

121 This is, of course, the allegiance of interest between the *Homo œconomicus* and liberal governmentality. Michel Foucault, *The Birth of Biopolitics: Lectures at the Collége de France, 1978–79*, ed. Michel Senellart, trans. Graham Burchell (Basingstoke, UK: Palgrave Macmillan, 2008).

122 For this line of critique, see Moore, *The Quantified Self in Precarity.*

123 For example, see Gerhard Dohrn-van Rossum, *History of the Hour: Clocks and Modern Temporal Orders* (Chicago: University of Chicago Press, 1996); Thompson, "Time, Work-Discipline, and Industrial Capitalism," *Past and Present* 38 (1967): 56–97.

124 Jenkins Jr., "Google and the Search for the Future."

125 For example, Amy Larocca, "The Wellness Epidemic," *Thecut*, June 27, 2017, https://www.thecut.com/2017/06/how-wellness-became-an-epidemic.html.

126 Joseph Dumit, *Drugs for Life: How Pharmaceutical Companies Define Our Health* (Durham, NC: Duke University Press, 2012).

127 N. Katherine Hayles, "Cognitive Assemblages: Technical Agency and Human Interactions," *Critical Inquiry* 43, no. 1 (2016): 43.

128 Danielle Keats Citron and Frank Pasquale, "The Scored Society: Due Process for Automated Predictions," *Washington Law Review* 89 (2014): 101–133.

129 For example, Justin Jouvenal, "The New Way Police Are Surveilling You: Calculating Your Threat 'Score,'" *The Washington Post*, January 10, 2016, https://www.washingtonpost.com/local/public-safety/the-new-way-police-are-surveilling-you-calculating-your-threat-score/2016/01/10/e42bccac-8e15-11e5-baf4-bdf37355da0c_story.html?utm_term=.5b5f41b1fe23.

130 Rob Horning, "Google Alert for the Soul," *New Inquiry*, April 12, 2013, http://thenewinquiry.com/essays/google-alert-for-the-soul/.

131 Sam Biddle, "Amazon's Home Surveillance Chief Declared War on 'Dirtbag Criminals' as Company Got Closer to Police," *The Intercept*, February 14, 2019, https://theintercept.com/2019/02/14/amazon-ring-police-surveillance/.

CHAPTER 5. BODIES INTO FACTS

1 Michael Milner, "Did Edward Snowden Tell Us Anything We Didn't Already Know?," *Chicago Reader*, June 25, 2013, http://www.chicagoreader.com/Bleader/archives/2013/06/25/did-edward-snowden-tell-us-anything-we-didnt-already-know.

2 Sarah Laskow, "A New Film Shows How Much We Knew, Pre-Snowden, about Internet Surveillance," *Columbia Journalism Review*, July 15, 2013, http://www.cjr.org/cloud_control/a_new_film_shows_exactly_how_m.php.

3 Glenn Greenwald, *No Place to Hide: Edward Snowden, the NSA and the Surveillance State* (London: Penguin Books, 2014), 19.

4 "We Already Knew the NSA Spies on Us. We Already Know Everything. Everything Is Boring," *Clickhole*, February 9, 2015, www.clickhole.com.

5 Quotations from, respectively, David Rowan, "Snowden: Big Revelations to Come, Reporting Them Is Not a Crime," *Wired*, March 18, 2014, http://www.wired.co.uk/news/archive/2014-03/18/snowden-ted; Eli Lake, "Spy Chief: We Should've Told You We Track Your Calls," *Daily Beast*, February 17, 2014, http://www.thedailybeast.com/articles/2014/02/17/spy-chief-we-should-ve-told-you-we-track-your-calls.html.

6 Brian Massumi, "Fear (The Spectrum Said)," *Positions* 13, no. 1 (2005): 35.

7 N. Katherine Hayles, *Unthought: The Power of the Cognitive Nonconscious* (Chicago: University of Chicago Press, 2017), 144.

8 See Jacques Derrida and Maurizio Ferraris, *A Taste for the Secret*, ed. Giacomo Donis and David Webb, trans. Giacomo Donis (Cambridge: Polity Press, 2001), 19–20.

9 Samuel R. Delany, "About Five Thousand One Hundred and Seventy Five Words," in *Sf: The Other Side of Realism*, ed. Thomas D. Clareson (Bowling Green, OH: Bowling Green University Popular Press, 1971), 10–11.

10 Tung-Hui Hu, *A Prehistory of the Cloud* (Cambridge, MA: MIT Press, 2015), ch. 1.

11 Colin Milburne, "{Zero Day} // Hacking as Applied Science Fiction" (presentation at the Department of History and Sociology of Science Fall 2015 Monday Workshop Series, Philadelphia, 2015).

12 Gabriella Coleman, *Hacker, Hoaxer, Whistleblower, Spy: The Many Faces of Anonymous* (London: Verso, 2014), 64.

13 Timothy Melley, *The Covert Sphere: Secrecy, Fiction, and the National Security State* (Ithaca, NY: Cornell University Press, 2012).

14 Roy Rappaport, *Ritual and Religion in the Making of Humanity* (Cambridge: Cambridge University Press, 1999); Victor Turner, "Liminal to Liminoid," in *Play, Flow, Ritual: An Essay in Comparative Symbology*, ed. Victor Turner (New York: Performing Arts Journal Publishing, 1982), 53–92.

15 Brian Knappenberger, "Why Care about the N.S.A.?," *The New York Times*, November 26, 2013, http://www.nytimes.com/video/opinion/100000002571435/why-care-about-the-nsa.html.

16 In 2008, the Foreign Intelligence Surveillance Court (FISC) granted the NSA permission to undertake particular kinds of upstream collection under section 702. This included "about collection"—that is, any communication that *mentioned* the targeted selector. Critics argued that this was typical of the NSA's tendency toward excessive surveillance; the agency insisted that given the nature of online communications and the surveillance systems, separating out a single individual from the data flow, at least not in this bulk collection stage, was impossible. The NSA ceased this particular form of collection in 2017, after regular self-reporting procedures to the FISC.

17 Crucially, the full targeting procedures—the guidelines analysts used to make best estimates on whether a selector should be targeted for collection—have not been released to the public. What we know is that "incidental collection," where Americans' information might be collected while collecting from targeted (foreign) selectors, is a regular part of the process. (This is distinct from "inadvertent collection," in which case a domestic selector was targeted in error and the collected information is required to be painstakingly removed.) Also see Jennifer Stisa Granick, *American Spies: Modern Surveillance, Why You Should Care, and What to Do about It* (Cambridge: Cambridge University Press, 2017), 106–15.

18 Granick, *American Spies*, ch. 7.

19 For an overview of the IMSI catcher, see Lisa Parks, "Rise of the IMSI Catcher," *Media Fields* 11 (2016).

20 As argued in chapter 3, this "everybody" rarely literally means everybody. In the context of post-9/11 counterterrorist efforts, there is a widespread tendency to overimagine the threat of the black, Muslim, male and mentally deviant. We might point to the New York Police Department's multiyear surveillance of Muslim neighborhoods and Muslim-owned shops that ultimately yielded no solid leads and two lawsuits against it. See Krishnadev Calamur, "NYPD Settles Pair of Lawsuits Over Muslim Surveillance," *The Atlantic*, January 7, 2016, http://www.theatlantic.com/national/archive/2016/01/nypd-surveillance-muslims-settlement/423174/; "AP's Probe into NYPD Intelligence Operations," Associated Press, 2012, http://www.ap.org/Index/AP-In-The-News/NYPD.

21 Micah Lee, "Edward Snowden Explains How To Reclaim Your Privacy," *The Intercept*, November 12, 2015, https://theintercept.com/2015/11/12/edward-snowden-explains-how-to-reclaim-your-privacy/.

22 "Operations Security (OPSEC)," US Department of Defense Education Activity. Accessed March 18, 2016, www.dodea.edu.

23 Hu, *A Prehistory of the Cloud*, 57.

24 Andy Greenberg, "How to Anonymize Everything You Do Online," *Wired*, June 17, 2014, http://www.wired.com/2014/06/be-anonymous-online/.

25 "Edward Snowden SXSW: Full Transcript and Video," *Inside.com*, March 10, 2014, http://blog.inside.com/blog/2014/3/10/edward-snowden-sxsw-full-transcription-and-video.

26 Ian Hacking, *The Taming of Chance* (Cambridge: Cambridge University Press, 1990), 160.

27 See Rappaport, *Ritual and Religion*, 118–23.

28 See Michael Betancourt, "The Demands of Agnotology::Surveillance," *Ctheory*, 2014, http://ctheory.net/ctheory_wp/the-demands-of-agnotologysurveillance.

29 Jodi Dean, "Publicity's Secret," *Political Theory* 29, no. 5 (2001): 625.

30 Dean, "Publicity's Secret," 642.

31 Wendy Hui Kyong Chun, *Updating to Remain the Same: Habitual New Media* (Cambridge, MA: MIT Press, 2015).

32 Quotations, respectively, Nina Eliasoph, *Avoiding Politics: How Americans Produce Apathy in Everyday* (Cambridge: Cambridge University Press, 1998), 134; Gijs van Oenen, "A Machine that Would Go of Itself: Interpassivity and Its Impact on Political Life," *Theory & Event* 9, no. 2 (2006), https://muse.jhu.edu/article/198813.

33 Michael Hayden, address to the National Press Club, Washington, DC, January 23, 2006.

34 Robert Pfaller, "Little Gestures of Disappearance(1) Interpassivity and the Theory of Ritual," *Journal of European Psychoanalysis* 16 (2003); Hagen Schölzel, "Beyond Interactivity. The Interpassive Hypotheses on 'Good Life' and Communication."; Gijs van Oenen, "Interpassivity Revisited: A Critical and Historical Reappraisal

of Interpassive Phenomena," *International Journal of Zizek Studies* 2, no. 2 (2002), https://muse.jhu.edu/article/198813.

35 Slavoj Žižek, "The Interpassive Subject," n.d., www.egs.edu.

36 Robert Pfaller, "Interpassivity and Misdemeanors. The Analysis of Ideology and the Zizekian Toolbox," *International Journal of Zizek Studies* 1, no. 1 (2001): 37.

37 Sun-ha Hong, "The Other-Publics: Mediated Othering and the Public Sphere in the Dreyfus Affair," *European Journal of Cultural Studies* 17, no. 6 (August 4, 2014): 665–681.

38 Slavoj Žižek, "The Inherent Transgression," *Cultural Values* 2, no. 1 (January 1998): 6–7.

39 Also see Svitlana Matviyenko, "Interpassive User: Complicity and the Returns of Cybernetics," *The Fibreculture Journal*, no. 25 (2015): 135–163.

40 Brian Knowlton, "Feinstein 'Open' to Hearings on Surveillance Programs," *The New York Times*, June 9, 2013, http://thecaucus.blogs.nytimes.com.

41 Melley, *The Covert Sphere*.

42 Justin Elliott and Theodoric Meyer, "Claim on 'Attacks Thwarted' by NSA Spreads Despite Lack of Evidence," *ProPublica*, October 23, 2013, http://www.propublica.org/article/claim-on-attacks-thwarted-by-nsa-spreads-despite-lack-of-evidence; Daniel Sterman, "Infographic: How the Government Exaggerated the Successes of NSA Surveillance," *Slate*, January 16, 2014, http://www.slate.com/blogs/future_tense/2014/01/16/nsa_surveillance_how_the_government_exaggerated_the_way_its_programs_stopped.html.

43 Julia Angwin, *Dragnet Nation: A Quest for Privacy, Security, and Freedom in a World of Relentless Surveillance* (New York: Times Books, 2014), chap. 3.

44 Richard A. Clarke, Michael J. Morrell, Geoggrey R. Stone, Cass R. Sunstein, and Peter Swire, "Liberty and Security in a Changing World," 2013, 119–120.

45 Robert O'Harrow Jr, "NSA Chief Asks a Skeptical Crowd of Hackers to Help Agency Do Its Job," *The Washington Post*, July 31, 2013, https://www.washingtonpost.com/world/national-security/nsa-chief-asks-a-skeptical-crowd-of-hackers-to-help-agency-do-its-job/2013/07/31/351096e4-fa15-11e2-8752-b41d7ed1f685_story.html.

46 "NSA Chief Keith Alexander Keynote @ Black Hat USA 2013 (w/ Slide Presenta-tion)," *Leaksource*, August 1, 2013. http://leaksource.info/2013/08/01/nsa-chief-keith-alexander-keynote-black-hat-usa-2013-w-slide-presentation/.

47 The NSA held "SIGINT 101 Seminars" for journalists in the early 2000s. One "course outline," first retrieved through a Freedom of Information Act by the *New York Sun*, describes the NSA's stated objective as *not* "about controlling what you write" but that the journalists "leave today with an understanding of [the NSA's] concern about 'the fact of' leaks—that is, the content of the material disclosed." Here, we find the same appeal to "real facts" as opposed to leaked ones. See Josh Gerstein, "Spies Prep Reporters on Protecting Secrets," *New York Sun*, September 27, 2007, http://www.nysun.com/national/spies-prep-reporters-on-protecting-secrets/63465/.

48 "Annual Open Hearing on Current and Projected National Security Threats to the United States," US Senate Select Committee on Intelligence, 113th Congress, Second Session, Washington, DC, 2014, www.intelligence.senate.gov; "Transcript: Senate Intelligence Hearing on National Security Threats," *The Washington Post*, January 29, 2014, https://www.washingtonpost.com/world/national-security/ transcript-senate-intelligence-hearing-on-national-security-threats/2014/01/29/ b5913184-8912-11e3-833c-33098f9e5267_story.html.

49 "Transcript: Senate Intelligence Hearing."

50 Peter Van Buren, "10 Myths about NSA Surveillance that Need Debunking," *MotherJones*, January 13, 2013, http://www.motherjones.com/politics/2014/01/10-myths-nsa-surveillance-debunk-edward-snowden-spying.

51 George Packer, "The Holder of Secrets," *The New Yorker*, October 20, 2014, http://www.newyorker.com/magazine/2014/10/20/holder-secrets.

52 Conor Friedersdorf, "New Surveillance Whistleblower: The NSA Violates the Constitution," *The Atlantic*, July 21, 2014, http://www.theatlantic.com/politics/archive/2014/07/a-new-surveillance-whistleblower-emerges/374722/.

53 Barry Siegel, "Judging State Secrets: Who Decides—and How?," in *After Snowden: Privacy, Secrecy, and Security in the Information Age*, ed. Ronald Goldfarb (New York: St. Martin's Press, 2015), 141–190.

54 Caren Bohan, "Lawmakers Urge Review of Domestic Spying, Patriot Act," *Chicago Tribune*, June 9, 2013, http://articles.chicagotribune.com/2013-06-09/news/sns-rt-us-usa-security-lawmakersbre9580ab-20130609_1_guardian-national-security-agency-surveillance.

55 James Risen and Laura Poitras, "N.S.A. Report Outlined Goals for More Power," *The New York Times*, November 22, 2013, http://www.nytimes.com/2013/11/23/us/politics/nsa-report-outlined-goals-for-more-power.html.

56 Melley, *The Covert Sphere*, 29.

57 Snowden, "Whistleblower Edward Snowden Gives 2013's Alternative Christmas Message," *Channel4.com*, December 25, 2013, www.channel4.com.

58 Jenny Hendrix, "NSA Surveillance Puts George Orwell's '1984' on Bestseller Lists," *Los Angeles Times*, June 11, 2013, http://articles.latimes.com/2013/jun/11/entertainment/la-et-jc-nsa-surveillance-puts-george-orwells-1984-on-bestseller-lists-20130611.

59 For example, Sarah Burris, "'Minority Report' Is Coming True: We Now Have Threat Scores to Match Our Credit Scores," *Los Angeles Times*, June 11, 2013, http://www.salon.com/2016/01/15/minority_report_is_coming_true_we_now_have_threat_scores_to_match_our_credit_scores_partner/; Greenwald, *No Place to Hide*.

60 Melley, *The Covert Sphere*, 115.

61 Richard Ledgett, "The NSA Responds to Edward Snowden's TED Talk," filmed March 20, 2014, at TED 2014, Vancouver, British Columbia, Canada, video, 33.19, http://www.ted.com/talks/richard_ledgett_the_nsa_responds_to_edward_snowdens_ted_talk.

62 Cory Doctorow, "Exclusive: Snowden Intelligence Docs Reveal UK Spooks' Malware Checklist," *Boingboing*, February 2, 2016, http://boingboing.net/2016/02/02/doxxing-sherlock-3.html.

63 Frank Lovece, "Soldier Showdown: Joe and Anthony Russo Take the Helm of 'Captain America' Franchise," *Filmjournal*, March 25, 2014, http://www.filmjournal.com/node/9232.

64 John Patterson, "How Hollywood Softened Us up for NSA Surveillance," *The Guardian*, June 16, 2013. http://www.theguardian.com/film/shortcuts/2013/jun/16/hollywood-softened-us-up-nsa-surveillance.

65 Joshua Rothman, "'Person of Interest': The TV Show that Predicted Edward Snowden," *The New Yorker*, January 14, 2014, http://www.newyorker.com/culture/culture-desk/person-of-interest-the-tv-show-that-predicted-edward-snowden.

66 Vicky Gan, "How TV's 'Person of Interest' Helps Us Understand the Surveillance Society," *Smithsonian.com*, October 24, 2013, http://www.smithsonianmag.com/smithsonian-institution/how-tvs-person-of-interest-helps-us-understand-the-surveillance-society-5407171/?no-ist.

67 Also see Hagen Schölzel, "Backing Away from Circles of Control: A Re-Reading of Interpassivity Theory's Perspectives on the Current Political Culture of Participation," *Empedocles: European Journal for the Philosophy of Communication* 8, no. 2 (2017): 187–203.

68 For an analogous case of institutional "cultures of disciplined improvisation" and their impact on working standards of truth and certainty, see Phaedra Daipha, *Masters of Uncertainty: Weather Forecasters and the Quest for Ground Truth* (Chicago: Chicago University Press, 2015).

69 For example, Il Chul Moon and Kathleen M. Carley, "Modeling and Simulating Terrorist Networks in Social and Geospatial Dimensions," *IEEE Intelligent Systems* 22, no. 5 (2007): 40–49.

70 For example, Peter Adey and Ben Anderson, "Anticipating Emergencies: Technologies of Preparedness and the Matter of Security," *Security Dialogue* 43, no. 2 (2012): 99–117.

71 See, respectively, Mark Andrejevic, "The Work of Watching One Another: Lateral Surveillance, Risk, and Governance," *Surveillance & Society* 2, no. 4 (2005): 479–497; Torin Monahan, *Surveillance in the Time of Insecurity* (New Brunswick, NJ: Rutgers University Press, 2010).

72 Amy Davidson, "Unclear Dangers," *The New Yorker*, May 18, 2015, http://www.newyorker.com/news/amy-davidson/the-n-s-a-s-spying-on-muslim-americans.

73 "Walker Leads Tightly Clustered GOP Field, Clinton Up Big Nationally," Public Policy Polling, May 13, 2015, https://www.publicpolicypolling.com/wp-content/uploads/2017/09/PPP_Release_National_51315.pdf.

74 Also see Peter Adey, "Facing Airport Security: Affect, Biopolitics, and the Preemptive Securitisation of the Mobile Body," *Environment and Planning D: Society and Space* 27, no. 2 (2009): 274–295; Claudia Aradau and Rens van Munster, "The Time/Space of Preparedness: Anticipating the 'Next Terrorist Attack,'" *European Journal*

of International Relations 13, no. 1 (March 1, 2007): 89–115; Ben Anderson, "Facing the Future Enemy: US Counterinsurgency Doctrine and the Pre-Insurgent," *Theory, Culture & Society* 28, no. 7–8 (2012): 216–240; Brian Massumi, "Potential Politics and the Primacy of Preemption," *Theory & Event* 10, no. 2 (July 19, 2007), https://muse.jhu.edu/issue/11693; Massumi, "The Future Birth of the Affective Fact: The Political Ontology of Threat," in *The Affect Theory Reader*, ed. Melissa Gregg and Gregory J. Seigworth (Durham, NC: Duke University Press, 2010), 52–70.

75 Trevor Aaronson, "The Sting: How the FBI Created a Terrorist," *The Intercept*, March 16, 2015, https://theintercept.com/2015/03/16/howthefbicreatedaterrorist/.

76 Peter Bergen, *United States of Jihad—Investigating America's Homegrown Terrorists* (New York: Crown Publishers, 2016).

77 For an analysis of the Newburgh Four in these lines, see Piotr M. Szpunar, "Pre-mediating Predisposition: Informants, Entrapment, and Connectivity in Counter-terrorism," *Critical Studies in Media Communication* 34, no. 4 (2017): 371–385.

78 Human Rights Watch, "Illusion of Justice: Human Rights Abuses in US Terrorism Prosecutions," 2014, https://www.hrw.org/sites/default/files/reports/usterror-ismo825_ForUpload_1_0.pdf.

79 Linda D. Kozaryn, "Alleged Al Qaeda 'Dirty Bomb' Operative in U.S. Military Custody," US Department of Defense, June 10, 2002, http://archive.defense.gov/news/newsarticle.aspx?id=43767.

80 They are primarily sourced from the Department of Justice prosecution lists of 2010, 2014, and 2015. Trevor Aaronson and Margot Williams, "Trial and Terror," *The Intercept*, accessed April 10, 2017, https://trial-and-terror.theintercept.com/.

81 Cassandra B. Carnright, "Affidavit, United States of America v. Joseph D. Jones and Edward Schimenti."

82 "Two Illinois Men Charged With Conspiring to Provide Material Support to ISIS," US Department of Justice, April 12, 2017, https://www.justice.gov/opa/pr/two-illinois-men-charged-conspiring-provide-material-support-isis.

83 Ellen Nakashima, "NSA Cites Case as Success of Phone Data-Collection Pro-gram," *The Washington Post*, August 8, 2013, https://www.washingtonpost.com/world/national-security/for-nsa-chief-terrorist-threat-drives-passion-to-collect-it-all/2013/07/14/3d26ef80-ea49-11e2-a301-ea5a8116d211_story.html.

84 Matthias Schwartz, "The Whole Haystack," *The New Yorker*, January 26, 2015, http://www.newyorker.com/magazine/2015/01/26/whole-haystack.

85 "Zero tolerance" interventions are part of a wider trend in threat management strategies. Increasingly, there is an effort to identify and mitigate uncertainties earlier and thus at a more amorphous stage. See, for instance, the distinction between "possible uncertainty" and "potential uncertainty" in Limor Samimian-Darash, "Governing Future Potential Biothreats," *Current Anthropology* 54, no. 1 (2013): 1–22.

86 Rebecca Lemov, "Guantanamo's Catch-22: The Uncertain Interrogation Subject," in *Modes of Uncertainty—Anthropological Cases*, ed. Limor Samimian-Darash and Paul Rabinow (Chicago: Chicago University Press, 2015), 88–104.

87 *Homegrown: The Counter-Terror Dilemma*, directed by Greg Baker, aired on February 8, 2016, on HBO.

88 Barker, *Homegrown*.

89 From a legal philosophy perspective, these dilemmas are part of the wider question of what presumption of innocence a citizen should enjoy in the age of dragnet surveillance. Are we to be guaranteed a general moral right to be trusted as good citizens? See Hadjimatheou, "Surveillance Technologies, Wrongful Criminalisation, and the Presumption of Innocence," *Philosophy and Technology* 30, no. 1 (2017): 39–54.

90 Barker, *Homegrown*. Also see Aradau and van Munster, "Governing Terrorism through Risk"; Melinda Cooper, "Pre-Empting Emergence: The Biological Turn in the War on Terror," *Theory, Culture & Society* 23, no. 4 (July 1, 2006): 113–135.

91 Jon Henley, "French Presidential Candidates Debate Key Election Issues—as It Happened," *The Guardian*, March 20, 2017, https://www.theguardian.com/world/live/2017/mar/20/french-presidential-candidates-debate-key-election-issues-live.

92 Chris Den Hond and Angela Charlton, "French Candidate Macron Wants to Fix Suburban Unrest," *Albuquerque Journal*, March 8, 2017, https://apnews.com/fbc-665e7cde842db95f8382abd0d81de.

93 Steven Erlanger, "After London Attack, Prime Minister Says, 'Enough Is Enough,'" *The New York Times*, June 4, 2017, https://www.nytimes.com/2017/06/04/world/europe/uk-london-attacks.html?_r=0.

94 See Russell J. Skiba, "Zero Tolerance, Zero Evidence: An Analysis of School Disciplinary Practice," Research Policy Report #SR2 (Bloomington: Indiana Education Policy Center, 2000); American Psychological Association Zero Tolerance Task Force, "Are Zero Tolerance Policies Effective in the Schools?: An Evidentiary Review and Recommendations," *American Psychologist* 63, no. 9 (2008): 852–862; Richard R. Verdugo, "Race-Ethnicity, Social Class, and Zero-Tolerance Policies," *Education and Urban Society* 35, no. 1 (2002): 50–75.

95 For example, Neil Smith, "Global Social Cleansing: Postliberal Revanchism and the Export of Zero Tolerance," *Social Justice* 28, no. 3 (2001): 68–74; John E. Eck and Edward R. Maguire, "Have Changes in Policing Reduced Violent Crime? An Assessment of the Evidence," in *The Crime Drop in America*, ed. Alfred Blumstein and Joel Wallman (Cambridge: Cambridge University Press, 2000), 207–265.

96 "Policy Strategy No. 5: Reclaiming the Public Spaces of New York," 1994.

97 James Bamford, *The Shadow Factory: The Ultra-Secret NSA from 9/11 to the Eavesdropping on America* (New York: Doubleday, 2008), 144.

98 For example, Derek Gregory, "From a View to a Kill: Drones and Late Modern War," *Theory, Culture & Society* 28, no. 7–8 (January 12, 2011): 188–215; Peter W. Singer, *Wired for War: The Robotics Revolution and Conflict in the 21st Century* (New York: Penguin, 2009).

99 Kelly A. Gates, *Our Biometric Future: Facial Recognition Technology and the Culture of Surveillance* (New York: New York University Press, 2011), 7, 48.

100 John Sifton, "A Brief History of Drones," *Nation*, February 7, 2012, http://www.the-nation.com/article/brief-history-drones/; Chris Woods, "The Story of America's

Very First Drone Strike," *The Atlantic*, May 30, 2015, http://www.theatlantic.com/international/archive/2015/05/america-first-drone-strike-afghanistan/394463/; Peter W. Singer, "Military Robots and the Future of War," *The New Atlantis* 23 (2009): 33–35.

101 See Peter M. Asaro, "The Labor of Surveillance and Bureaucratized Killing: New Subjectivities of Military Drone Operators," *Social Semiotics* 23, no. 2 (2013): 196–224.

102 Ryan Devereaux, "Manhunting in the Hindu Kush," *The Intercept*, October 15, 2015, https://theintercept.com/drone-papers/manhunting-in-the-hindu-kush/.

103 Jo Becker and Scott Shane, "Secret 'Kill List' Proves a Test of Obama's Principles and Will," *The New York Times*, May 29, 2012, http://www.nytimes.com/2012/05/29/world/obamas-leadership-in-war-on-al-qaeda.html?pagewanted=1&_r=1; Christopher Daase and Oliver Kessler, "Knowns and Unknowns in the 'War on Terror': Uncertainty and the Political Construction of Danger," *Security Dialogue* 38, no. 4 (2007): 420.

104 See Lauren Wilcox, "Embodying Algorithmic War: Gender, Race, and the Posthuman in Drone Warfare," *Security Dialogue* 48, no. 1 (2016): 19–20.

105 Schwartz, "The Whole Haystack."

106 Mark Andrejevic, "To Preempt a Thief," *International Journal of Communication* 11 (2017): 879–896.

107 Gregory, "From a View to a Kill," 196.

108 Richard Barrett, "Don't Turn Security into Theater," CNN, May 6, 2013, http://globalpublicsquare.blogs.cnn.com/2013/05/06/dont-turn-security-into-theater/.

109 Debbie Vickers, producer, and Jay Leno, *The Tonight Show with Jay Leno*, episode aired August 6, 2013, on NBC.

110 Jeffrey Goldberg, "The Obama Doctrine," *The Atlantic*, March 10, 2016, http://www.theatlantic.com/magazine/archive/2016/04/the-obama-doctrine/471525/#article-comments.

111 Andrew Shaver, "You're More Likely to Be Fatally Crushed by Furniture than Killed by a Terrorist," *The Washington Post*, November 23, 2015, https://www.washingtonpost.com/news/monkey-cage/wp/2015/11/23/youre-more-likely-to-be-fatally-crushed-by-furniture-than-killed-by-a-terrorist/.

112 Michael Hayden, "Statement for the Record," October 17, 2002.

113 Respectively, Michael Hayden, *Playing to the Edge: American Intelligence in the Age of Terror*; Hayden, address to the National Press Club, Washington, DC, January 23, 2006.

114 Obama, "Transcript: Obama's Remarks on NSA Controversy," *Wall Street Journal*, March 14, 2013, http://blogs.wsj.com/washwire/2013/06/07/transcript-what-obama-said-on-nsa-controversy/.

115 Schwartz, "The Whole Haystack."

116 Harvey Luskin Molotch, *Against Security: How We Go Wrong at Airports, Subways, and Other Sites of Ambiguous Danger* (New Brunswick, NJ: Rutgers University Press, 2010), 54.

117 Frank H. Knight, *Risk, Uncertainty and Profit* (New York: Augustus M. Kelley, 1964). Also see Paul Sollie, "Ethics, Technology Development and Uncertainty: An Outline for Any Future Ethics of Technology," *Journal of Information, Communication and Ethics in Society* 5, no. 4 (2007): 293–306.

118 See Limor Samimian-Darash and Paul Rabinow, "Introduction," in *Modes of Uncertainty—Anthropological Cases*, ed. Limor Samimian-Darash and Paul Rabinow (Chicago: Chicago University Press, 2015), 1–9.

119 Robert N. Procter, "Agnotology: A Missing Term to Describe the Cultural Production of Ignorance (and Its Study)," in *Agnotology: The Making & Unmaking of Ignorance*, ed. Robert N. Procter and Londa Schiebinger (Stanford, CA: Stanford University Press, 2008), 1–33.

120 For an overview, see Pat O'Malley, *Risk, Uncertainty and Government* (London: Glasshouse Press, 2004).

121 Nikolas Rose, "At Risk of Madness," in *Embracing Risk: The Changing Culture of Insurance and Responsibility*, ed. Tom Baker and Jonathan Simon (Chicago: Chicago University Press, 2002), 209–237; Nikolas Rose, Pat O'Malley, and Mariana Valverde, "Governmentality," *Annual Review of Law and Social Science* 2, no. 1 (2006): 83–104.

122 Ulrich Beck, *Risk Society: Towards a New Modernity* (London: Sage, 1992).

123 Ulrich Beck, *World at Risk* (Malden, MA: Polity Press, 2009).

124 Mitchell Dean, "Risk, Calculable and Incalculable," *Soziale Welt* 49 (1998): 25–42.

125 Hacking, *The Taming of Chance*, 97.

126 Allan Sekula, "The Body and the Archive," *October* 39 (1986): 3–64; Stephan Jay Gould, *The Mismeasure of Man: The Definitive Refutation to the Argument of The Bell Curve* (New York: W. W. Norton, 1996).

127 Hacking, *The Taming of Chance*.

128 Catherine E. Althaus, "A Disciplinary Perspective on the Epistemological Status of Risk," *Risk Analysis* 25, no. 3 (2005): 570.

129 See Louise Amoore, "Data Derivatives: On the Emergence of a Security Risk Calculus for Our Times," *Theory, Culture & Society* 28, no. 6 (2011): 24–43; Massumi, "Potential Politics and the Primacy of Preemption"; Massumi, "The Future Birth of the Affective Fact."

130 See Aradau and van Munster, "Governing Terrorism through Risk."

131 Matthew G. Hannah, "(Mis)Adventures in Rumsfeld Space," *GeoJournal* 75, no. 4 (2010): 397–406. Also see Bruce Schneier, "It's Smart Politics to Exaggerate Terrorist Threats," CNN, May 20, 2013, https://www.cnn.com/2013/05/20/opinion/schneier-security-politics/index.html; Daniel Moeckli, *Human Rights and Non-Discrimination in the "War on Terror"* (Oxford: Oxford University Press, 2008), 40.

132 Tony Blair, "Full Text: Tony Blair's Speech," *The Guardian*, March 5, 2004, www.theguardian.com.

133 Respectively, Emmeline Taylor, *Surveillance Schools: Security, Discipline and Control in Contemporary Education* (Basingstoke, UK: Palgrave Macmillan, 2013), 19; Schneier, "It's Smart Politics."

134 Daase and Kessler, "Knowns and Unknowns in the 'War on Terror.'"
135 Massumi, "The Future Birth of the Accective Fact."
136 From Mark Andrejevic, *InfoGlut: How Too Much Information Is Changing the Way We Think and Know* (New York: Routledge, 2013), 16–17.
137 The term was popularized during the 2010s through journalists such as James Fallows at *The Atlantic*, for example, James Fallows, "Truth, Lies, Politics, and the Press, in Three Acts," *The Atlantic*, August 18, 2012, https://www.theatlantic.com/politics/archive/2012/08/truth-lies-politics-and-the-press-in-three-acts/261297/; Jill Lepore, "After the Fact," *The New Yorker*, March 21, 2016, http://www.newyorker.com/magazine/2016/03/21/the-internet-of-us-and-the-end-of-facts?mbid=social_facebook.

CHAPTER 6. DATA-SENSE AND NON-SENSE

1 Epigraph from Lorraine J. Daston and Peter Galison, *Objectivity* (New York: Zone Books, 2007), 372.
2 Also see Ed Finn, *What Algorithms Want: Imagination in the Age of Computing* (Cambridge, MA: MIT Press, 2017).
3 Friedrich A. Kittler, *Gramophone, Film, Typewriter*, tr. Geoffrey Winthrop-Young and Michael Wutz (Stanford, CA: Stanford University Press, 1986).
4 David Weinberger, "Our Machines Now Have Knowledge We'll Never Understand," *Wired*, April 18, 2017, https://www.wired.com/story/our-machines-now-have-knowledge-well-never-understand/.
5 These dilemmas around nonhuman perception predate the internet by far. For analyses of early film and early photography, see, respectively, Harun Farocki, "Phantom Images," *Public* 29 (2004): 13; Joanna Zylinska, *Nonhuman Photography* (Cambridge, MA: MIT Press, 2017), 20–21.
6 Trevor Paglen, "Invisible Images (Your Pictures Are Looking at You)," *New Inquiry*, December 8, 2016, https://thenewinquiry.com/invisible-images-your-pictures-are-looking-at-you/.
7 Here, I am using Donald Landes's translation of Merleau-Ponty, *Phenomenology of Perception* (London: Routledge, 2012).
8 Ian Bogost, "Why Nothing Works Anymore," *The Atlantic*, February 23, 2017, https://www.theatlantic.com/technology/archive/2017/02/the-singularity-in-the-toilet-stall/517551/.
9 Tanya Nyong'o, "Plenary 4" (presentation at the Affect Theory Conference | Worldings | Tensions | Futures in Lancaster, PA, October 14–17, 2015).
10 Others have used "data-sense" to describe how human sensemaking interfaces with machine-produced data and the machines as physical objects. See Deborah Lupton, Christine Heyes Labond, and Shanti Sumartojo, "Personal Data Contexts, Data Sense, and Self-Tracking Cycling," *International Journal of Communication* 12 (2018): 647–666.
11 Gary Wolf and Kevin Kelly, "Wired's Gary Wolf & Kevin Kelly Talk the Quantified Self" (presentation at the WIRED Health Conference: Living by Numbers, New

York, October 15–16, 2012). Also see Pete Mortensen, "The Future Of Technology Isn't Mobile, It's Contextual," *Fast Company*, May 24, 2013, http://www.fastcode-sign.com/1672531/the-future-of-technology-isnt-mobile-its-contextual.

12 Wolf and Kelly, "Wired's Gary Wolf & Kevin Kelly."

13 Phoebe V. Moore, *The Quantified Self in Precarity—Work, Technology and What Counts* (Abingdon, UK: Routledge, 2018).

14 Also see Tamar Sharon and Dorien Zandbergen, "From Data Fetishism to Quantifying Selves: Self-Tracking Practices and the Other Values of Data," *New Media & Society* 19, no. 11 (2017): 1700.

15 Mette Dyhrberg, interview, May 26, 2015.

16 Ernesto Ramirez, "Larry Smarr: Where There Is Data There Is Hope," *Quantified Self*, February 12, 2013, http://quantifiedself.com/2013/02/larry_smarr_crone-shope_in_data/; Jon Cohen, "The Patient of the Future," *MIT Technology Review*, February 21, 2012. https://www.technologyreview.com/s/426968/the-patient-of-the-future/.

17 Mark Bowden, "The Measured Man," *The Atlantic* (July/August 2012), http://www.theatlantic.com/magazine/archive/2012/07/the-measured-man/309018/.

18 For a different take on self-tracking's cultural affinity to posthumanism, see Stefan Danter, Ulfried Reichardt, and Regina Schober, "Theorising the Quantified Self and Posthumanist Agency," *Digital Culture & Society* 2, no. 1 (2016): P53–67.

19 See Melanie Swan, "The Quantified Self: Fundamental Disruption in Big Data Science and Biological Discovery," *Big Data* 1, no. 2 (2013): 95.

20 See N. Katherine Hayles, *How We Became Posthuman: Virtual Bodies in Cybernetics, Literature, and Informatics* (Chicago: University of Chicago Press, 1999); Deborah Lupton, "Self-Tracking Cultures: Towards a Sociology of Personal Informatics" (presentation at the OzCHI '14: Proceedings of the 26th Australian Computer-Human Interaction Conference: Designing Futures, the Future of Design, Sydney, December 2–5, 2014).

21 Fred Turner, *From Counterculture to Cyberculture: Stewart Brand, the Whole Earth Network, and the Rise of Digital Utopianism* (Chicago: University of Chicago Press, 2006).

22 N. Katherine Hayles, "The Cognitive Nonconscious: Enlarging the Mind of the Humanities," *Critical Inquiry* 42 (2016): 784; N. Katherine Hayles, *Unthought: The Power of the Cognitive Nonconscious* (Chicago: University of Chicago Press, 2017), 6.

23 For a similarly motivated critique of experience, see Josh Berson, *Computable Bodies: Instrumented Life and the Human Somatic Niche* (London: Bloomsbury Academic, 2015).

24 Hayles, *Unthought*.

25 Hayles, *Unthought*, 22–23, 50–55.

26 This process has alternatively been described as "biopedagogy." Aristea Fotopou-lou and Kate O'Riordan, "Biosensory Experiences and Media Materiality Fitbit and Biosensors: Imaginaries and Material Instantiations" (presentation at IR16, Phoenix, AZ, October 21–24, 2015).

27 Merleau-Ponty, *Phenomenology of Perception*.

28 Also see Berson, *Computable Bodies*, 39.

29 Robert R. Morris, "A Shocking Solution to Facebook Addiction," *Medium*, August 21, 2013, https://medium.com/@robertrmorris/a-shocking-solution-to-facebook-addiction-d1f5a14e2943.

30 Maneesh Sethi, "Why I Hired a Girl on Craigslist to Slap Me in the Face—And How It Quadrupled My Productivity," *Hack the System*, October 16, 2012, http://hackthesystem.com/blog/why-i-hired-a-girl-on-craigslist-to-slap-me-in-the-face-and-why-it-quadrupled-my-productivity/.

31 Also see Mark B. N. Hansen, "Engineering Pre-Individual Potentiality: Technics, Transindividuation, and 21st-Century Media," *SubStance* 41, no. 3 (2012): 32–59.

32 Kevin Kelly, "Self-Tracking? You Will," KK, March 25, 2011, http://kk.org/thetechnium/self-tracking-y/. Also see Deborah Lupton, "Understanding the Human Machine," *IEEE Technology and Society Magazine* 32, no. 4 (2013): 25–30.

33 For a similar notion of the "technological default," see Belinda Barnet, "The Secret Life of Our Prostheses," *Ctheory* Theorizing (2015).

34 Jack Bratich, "Public Secrecy and Immanent Security," *Cultural Studies* 20, no. 4–5 (2006): 493.

35 Leo Marx, "Technology: The Emergence of a Hazardous Concept," *Technology and Culture* 51, no. 3 (2010): 561–577.

36 Jennifer A. Chandler, "'Obligatory Technologies': Explaining Why People Feel Compelled to Use Certain Technologies," *Bulletin of Science, Technology & Society* 32, no. 4 (2012): 255–264.

37 Peter E. S. Freund, "Civilised Bodies Redux: Seams in the Cyborg," *Social Theory & Health* 2, no. 3 (2004): 273–289.

38 Tung-Hui Hu, *A Prehistory of the Cloud* (Cambridge, MA: MIT Press, 2015).

39 Gary Wolf, interview, August 8, 2015.

40 Gary Wolf, "The Data-Driven Life," *The New York Times*, April 28, 2010, http://www.nytimes.com/2010/05/02/magazine/02self-measurement-t.html?_r=1.

41 Daston and Galison, *Objectivity*, 174–76, 203.

42 Theodore M. Porter, "The Objective Self," *Victorian Studies* 50, no. 4 (2014): 641–647.

43 Lorraine J. Daston, "Objectivity and the Escape from Perspective," *Social Studies of Science* 22 (1992): 597–618; Steve Shapin, *The Scientific Life: A Moral History of a Late Modern Vocation* (Chicago: University of Chicago Press, 2008).

44 Marx, "Technology."

45 Landes, *The Unbound Prometheus: Technological Change and Industrial Development in Western Europe from 1750 to the Present* (Cambridge: Cambridge University Press, 1969).

46 Leo Marx, "The Idea of 'Technology' and Postmodern Pessimism," in *Does Technology Drive History? The Dilemma of Technological Determinism*, ed. Merritt Roe Smith and Leo Marx (Cambridge, MA: MIT Press, 1994), 237–258; Marx, "Technology."

47 For example, Langdon Winner, *Autonomous Technology: Technics-out-of-Control as a Theme in Political Thought* (Cambridge, MA: MIT Press, 1977); Lewis Mumford, *Technics and Civilization* (New York: Harcourt, Brace and World, 1963); Jacques Ellul, *The Technological Society* (New York: Knopf, 1964).

48 John Bagnell Bury, *The Idea of Progress: An Inquiry into Its Origin and Growth* (London: Macmillan, 1920); van Lente, "Forceful Futures: From Promise to Requirement," in *Contested Futures: A Sociology of Prospective Techno-Science*, ed. Nik Brown, Brian Rappert, and Andrew Webster (Aldershot, UK: Ashgate, 2000), 43–64.

49 For a different take on a similar problem, see Jasanoff, "Future Imperfect: Science, Technology, and the Imaginations of Modernity," in *Dreamscapes of Modernity: Sociotechnical Imaginaries and the Fabrication of Power*, ed. Sheila Jasanoff and Sang-Hyun Kim (Chicago: Chicago University Press, 2015), 1–33.

50 Michael L. Smith, "Recourse of Empire: Landscapes of Progress in Technological America," in *Does Technology Drive History?: The Dilemma of Technological Determinism*, ed. Merritt Roe Smith and Leo Marx (Cambridge, MA: MIT Press, 1994), 40.

51 See David Edgerton, "Innovation, Technology, or History: What Is the Historiography of Technology about?," *Technology and Culture* 51, no. 3 (2010): 680–697; Kline, "Technological Determinism," in *International Encyclopedia of the Social and Behavioral Sciences* (New York: Elsevier, 2001), 15495–15498.

52 For example, Tarleton Gillespie, "The Politics of 'Platforms,'" *New Media & Society* 12, no. 3 (2010): 347–364.

53 For example, José van Dijck, *The Culture of Connectivity: A Critical History of Social Media* (Cambridge: Oxford University Press, 2013).

54 Howard Rheingold, *The Virtual Community: Homesteading on the Electronic Frontier* (Cambridge, MA: MIT Press, 2000).

55 Hu, *A Prehistory of the Cloud*, xxiv.

56 Hu, *A Prehistory of the Cloud*, xvii.

57 Sara M. Watson, "Data Is the New '___,'" *dis magazine*, 2015, http://dismagazine.com.

58 Also see Hito Steyerl, "A Sea of Data: Apophenia and Pattern (Mis-)Recognition," *E-Flux* 72 (2016).

59 The "soft biopolitics" of John Cheney-Lippold, *We Are Data—Algorithms and the Making of Our Digital Selves* (New York: New York University Press, 2017). Similar arguments about datafication's modality of control can be found across works such as Cathy O'Neill, *Weapons of Math Destruction: How Big Data Increases Inequality and Threatens Democracy* (New York: Crown, 2016) and Finn, *What Algorithms Want*.

60 For example, Michel Foucault, *The Birth of Biopolitics: Lectures at the Collége de France, 1978–79*, ed. Michel Senellart, trans. Graham Burchell (Basingstoke, UK: Palgrave Macmillan, 2008).

61 Gilles Deleuze, "Postscript on the Societies of Control," *October* 59 (1992): 3–7.

62 Foucault had emphasized that sovereign and disciplinary forms of power both endure well into modern society.

63 Also see James Ash, *Phase Media: Space, Time and the Politics of Smart Objects* (New York: Bloomsbury Academic, 2018), 124–25.

64 Used in a different but resonant context by Lemov, *Database of Dreams: The Lost Quest to Catalog Humanity* (New Haven, CT: Yale University Press, 2015), 250.

65 Natasha Dow Schüll, "Data for Life: Wearable Technology and the Design of Self-Care," *Biosocieties* 11, no. 3 (2016): 317–333.

66 See Natasha Dow Schüll, "Self-Tracking Technology from Compass to Thermostat" (presentation at Streams of Consciousness conference, University of Warwick, Warwick, UK, April 21–22, 2016).

67 Scott Dadich, "Science Fiction Helps Make Sense of an Uncertain Future," *Wired*, 3, December 13, 2016. https://www.wired.com/2016/12/editors-letter-january-2017/.

68 Charlie Jane Anders, "Stochastic Fancy," *Wired*, January 2017, https://www.wired.com/2016/1/2/charlie-jane-anders-robot-therapist/.

69 At the level of the everyday, many of these attitudes, and the predicaments they are born out of, echo Lauren Berlant's work on coping and bearing, for example, Berlant, *Cruel Optimism* (Durham, NC: Duke University Press, 2011).

70 Marcel Mauss in John Durham Peters, *The Marvelous Clouds: Toward a Philosophy of Elemental Media* (Chicago: University of Chicago Press, 2015), 80–81.

CONCLUSION

1 Epigraph from Joseph Weizenbaum, *Computer Power and Human Reason: From Judgment to Calculation* (New York: W. H. Freeman, 1976), x.

2 For the link between ubiquitous computing and the idea of paternalism, see Sarah Spiekermann and Frank Pallas, "Technology Paternalism—Wider Implications of Ubiquitous Computing," *Poiesis Und Praxis* 4, no. 1 (2006): 6–18.

3 Samuel Sinyangwe in Adam Foss, Samuel Sinyangwe, Julia Angwin, Charmaine Arthur, and Kim Foxx, "Automating (In)Justice: Policing and Sentencing in the Algorithm Age" (presentation at Data 4 Black Lives, Cambridge, MA, October 17–19, 2017).

4 See the argument in Nabil Hassein, "Against Black Inclusion in Facial Recognition," *Digital Talking Drum*, August 15, 2017, https://digitaltalkingdrum.com/2017/08/15/against-black-inclusion-in-facial-recognition/.

5 James C. Scott, *Seeing like a State: How Certain Schemes to Improve the Human Condition Have Failed* (New Haven, CT: Yale University Press, 1998).

6 Scott, *Seeing like a State*, 11.

7 Scott, *Seeing like a State*, 93.

8 Martin Heidegger, "The Question Concerning Technology," 1949, http://www.wright.edu/cola/Dept/PHL/Class/P.Internet/PITexts/QCT.html.

9 Mark Warschauer and Morgan Ames, "Can One Laptop Per Child Save the World's Poor?," *Journal of International Affairs* 64, no. 1 (2010): 35.

10 Frank Pasquale, "Machine Learning, Meaning, & Law" (presentation at the Cybernetics Conference, New York, November 15–17, 2017).

11 A point made in the more specific case of new military technology by Peter W. Singer, "The Ethics of Killer Applications: Why Is It So Hard To Talk About Morality When It Comes to New Military Technology?," *Journal of Military Ethics* 9, no. 4 (2010): 299–312. Also see Armin Grunwald, "Against Over-Estimating the Role of Ethics in Technology Development," *Science and Engineering Ethics* 6 (2000): 181–196.

12 Harro van Lente, "Forceful Futures: From Promise to Requirement," In *Contested Futures: A Sociology of Prospective Techno-Science*, ed. Nik Brown, Brian Rappert, and Andrew Webster (Aldershot, UK: Ashgate, 2000), 43–64.

13 Eric Cohen, *In the Shadow of Progress: Being Human in the Age of Technology* (New York: New Atlantis Books, 2008).

14 For example, Sheila Jasanoff, *The Ethics of Invention: Technology and the Human Future* (New York: W. W. Norton, 2016).

15 For example, Yale Brozen, "The Value of Technological Change," *Ethics* 62, no. 4 (1952): 249–265; Karl Lautenschlager, "Controlling Military Technology," *Ethics* 95, no. 3 (1985): 692–711.

16 Also see David Golumbia, *The Cultural Logic of Computation* (Cambridge, MA: Harvard University Press, 2009).

17 For an overview, see Brian Leiter, "The Hermeneutics of Suspicion: Recovering Marx, Nietzsche, and Freud," in *The Future for Philosophy*, ed. Brian Leiter (Oxford: Clarendon Press, 2004), 74–105.

18 For example, James H. Moor, "What Is Computer Ethics?," *Metaphilosophy* 16, no. 4 (1985): 266–275.

19 Wendy Hui Kyong Chun, *Updating to Remain the Same: Habitual New Media* (Cambridge, MA: MIT Press, 2015).

20 For example, Thomas Douglas, "Human Enhancement and Supra-Personal Moral Status," *Philosophical Studies* 162, no. 3 (2013): 473–497; Moor, "The Nature, Importance, and Difficulty of Machine Ethics," in *Machine Ethics*, ed. Michael Anderson and Susan Leigh Anderson (Cambridge: Cambridge University Press, 2011), 13–20. For the problem of responsibility allocation, see Luciano Floridi, "Robots, Jobs, Taxes, and Responsibilities," *Philosophy and Technology* 30, no. 1 (2017): 1–4.

21 Mark Coeckelbergh, "Is Ethics of Robotics about Robots? Philosophy of Robotics beyond Realism and Individualism," *Law, Innovation and Technology* 3, no. 2 (2011): 241–250.

22 See David J. Gunkel, *The Machine Question: Critical Perspectives on AI, Robots, and Ethics* (Cambridge, MA: MIT Press, 2012), 159.

23 As parsed in Lorenzeo Magnani, *Morality in a Technological World: Knowledge as Duty* (Cambridge: Cambridge University Press, 2007).

24 Paul Sollie, "Ethics, Technology Development and Uncertainty: An Outline for Any Future Ethics of Technology," *Journal of Information, Communication and Ethics in Society* 5, no. 4 (2007): 293–306.

25 For example, Christian Munthe, *The Price of Precaution and the Ethics of Risk* (Dordrecht: Springer, 2011).

26 Sven Ove Hansson, "From the Casino to the Jungle: Dealing with Uncertainty in Technological Risk Management," *Synthese* 168, no. 3 (2009): 423–432.

27 Also see Sven Ove Hansson, "Philosophical Perspectives on Risk" (presentation at Research in Ethics and Engineering, Delft, Netherlands, 2002).

28 Shannon Vallor, *Technology and the Virtues: A Philosophical Guide to a Future Worth Wanting* (Oxford: Oxford University Press, 2016).

29 Nora Bateson, "Warm Data: Contextual Research and New Forms of Information," *Hackernoon*, May 28, 2017, https://hackernoon.com/warm-data-9fofcd2a828c.

30 It should be clear that these questions are raised with, rather than against, Vallor's work and the turn to virtue ethics.

31 Bruno Latour, "On Technical Mediation—Philosophy, Sociology, Genealogy," *Common Knowledge* 3, no. 2 (1994): 38. In this vein, also see Bruno Latour, "Morality and Technology: The End of the Means," *Theory, Culture & Society* 19, no. 5–6 (2002): 247–260.

32 Also see Frank Pasquale, "Odd Numbers," *Real Life*, August 20, 2018, http://reallifemag.com/odd-numbers/.

33 Latour, "Morality and Technology."

34 Langdon Winner, "Do Artifacts Have Politics?," *Daedalus* 109, no. 1 (1980): 121.

35 Tung-Hui Hu, *A Prehistory of the Cloud* (Cambridge, MA: MIT Press, 2015).

36 For example, Holland, "Finally, a Better Way for 'Quantified Self' Products to Collect Personal Data," *Fast Company*, October 18, 2013, http://www.fastcompany.com/3020212/finally-a-better-way-for-quantified-self-products-to-collect-personal-data; Joshua Reeves, "Frictionless Tracking with Beeminder Autodata" (presentation at the Quantified Self 2015 Conference, San Francisco, June 18–20, 2015).

37 Carolyn Marvin, *When Old Technologies Were New: Thinking About Electric Communication in the Late Nineteenth Century* (Oxford: Oxford University Press, 1988).

38 A similar point is made regarding the early days of computing technologies in Weizenbaum, *Computer Power and Human Reason*, 28–31.

39 Antoinette Rouvroy, "Privacy, Data Protection, and the Unprecedented Challenges of Ambient Intelligence," *Studies in Ethics, Law, and Technology* 2, no. 1 (2008): 15.

40 Sarah E. Igo, *The Known Citizen: A History of Privacy in Modern America* (Cambridge, MA: Harvard University Press, 2018).

41 Ian Bogost, "Why Nothing Works Anymore," *The Atlantic*, February 23, 2017, https://www.theatlantic.com/technology/archive/2017/02/the-singularity-in-the-toilet-stall/517551/.

42 An actual example in the case of a Japanese subway company. See Mark Gawne, "The Modulation and Ordering of Affect: From Emotion Recognition Technology to the Critique of Class Composition," *Fibreculture* 21 (2012): 98–123.

43 Hacking, *The Taming of Chance* (Cambridge: Cambridge University Press, 1990), 145.

44 Borgmann described this as the distinction between 'commanding' and 'disposable' material objects. See Albert Borgmann, "The Moral Significance of the Material Culture," *Inquiry: An Interdisciplinary Journal of Philosophy* 35, no. 3–4 (1992): 291–300.

45 Simone de Beauvoir, *The Ethics of Ambiguity* (New York: Philosophical Library, 1948), 85.

BIBLIOGRAPHY

Aaronson, Trevor. "The Sting: How the FBI Created a Terrorist." *The Intercept*, March 16, 2015. https://theintercept.com/2015/03/16/howthefbicreatedaterrorist/.

———, and Margot Williams. "Trial and Terror." *The Intercept*. Accessed April 10, 2017. https://trial-and-terror.theintercept.com/.

Ackerman, Linda. "Mobile Health and Fitness Applications and Information Privacy," 2013. San Diego, CA: Privacy Rights Clearinghouse.

Adey, Peter. "Facing Airport Security: Affect, Biopolitics, and the Preemptive Securitisation of the Mobile Body." *Environment and Planning D: Society and Space* 27, no. 2 (2009): 274–295.

———, and Ben Anderson. "Anticipating Emergencies: Technologies of Preparedness and the Matter of Security." *Security Dialogue* 43, no. 2 (2012): 99–117.

Almasy, Steve. "Journalist: Snowden Has More Documents that Could Harm U.S." CNN, July 15, 2013. www.cnn.com/2013/07/14/politics/nsa-leak-greenwald/.

Althaus, Catherine E. "A Disciplinary Perspective on the Epistemological Status of Risk." *Risk Analysis* 25, no. 3 (2005): 567–588.

American Psychological Association Zero Tolerance Task Force. "Are Zero Tolerance Policies Effective in the Schools?: An Evidentiary Review and Recommendations." *American Psychologist* 63, no. 9 (2008): 852–862.

Amoore, Louise. "Data Derivatives: On the Emergence of a Security Risk Calculus for Our Times." *Theory, Culture & Society* 28, no. 6 (2011): 24–43.

———. "Doubt and the Algorithm: On the Partial Accounts of Machine Learning." *Theory, Culture & Society* (2019).

Anders, Charlie Jane. "Stochastic Fancy." *Wired*, January 2017. https://www.wired.com/2016/1/2/charlie-jane-anders-robot-therapist/.

Anderson, Ben. "Affect and Biopower: Towards a Politics of Life." *Transactions of the Institute of British Geographers* 37, no. 1 (2012): 28–43.

———. "Facing the Future Enemy: US Counterinsurgency Doctrine and the Pre-Insurgent." *Theory, Culture & Society* 28, no. 7–8 (2012): 216–240.

Andrejevic, Mark. "The Work of Watching One Another: Lateral Surveillance, Risk, and Governance." *Surveillance & Society* 2, no. 4 (2005): 479–497.

———. *InfoGlut: How Too Much Information Is Changing the Way We Think and Know.* New York: Routledge, 2013.

———. "To Preempt a Thief." *International Journal of Communication* 11 (2017): 879–896.

———, and Mark Burdon. "Defining the Sensor Society." *Television & New Media* 16, no. 1 (2015): 19–36.

Angwin, Julia. *Dragnet Nation: A Quest for Privacy, Security, and Freedom in a World of Relentless Surveillance.* New York: Times Books, 2014.

Anti-Defamation League. "Tom Metzger." Accessed February 20, 2016. http://archive.adl.org/learn/ext_us/tom-metzger/ideology.html?LEARN_Cat=Extremism&LEARN_SubCat=Extremism_in_America&xpicked=2&item=7.

Aradau, Claudia, and Tobias Blanke. "The (Big) Data-Security Assemblage: Knowledge and Critique." *Big Data & Society* 2, no. 2 (2015): 1–12.

Aradau, Claudia, and Rens van Munster. "Governing Terrorism through Risk: Taking Precautions, (Un)Knowing the Future." *European Journal of International Relations* 13, no. 1 (March 1, 2007): 89–115.

———. "The Time/Space of Preparedness: Anticipating the 'Next Terrorist Attack.'" *Space and Culture* 15, no. 2 (March 21, 2012): 98–109.

Aryan Vanguard. "Lone Wolf Tactical Concept." June 2012. http://aryanvanguard.blogspot.com/2012/06/lone-wolf-tactical-concept.html.

Asaro, Peter M. "The Labor of Surveillance and Bureaucratized Killing: New Subjectivities of Military Drone Operators." *Social Semiotics* 23, no. 2 (2013): 196–224.

Ash, James. *Phase Media: Space, Time and the Politics of Smart Objects.* New York: Bloomsbury Academic, 2018.

Associated Press. "AP's Probe into NYPD Intelligence Operations." 2012. http://www.ap.org/Index/AP-In-The-News/NYPD.

Aupers, Stef. "'Trust No One': Modernization, Paranoia and Conspiracy Culture." *European Journal of Communication* 27, no. 1 (2012): 22–34.

Bakker, Edwin, and Beatrice de Graaf. "Lone Wolves: How to Prevent This Phenomenon?" Presentation at *Expert Meeting Lone Wolves*, The Hague, the Netherlands, November 5, 2010. http://ctheory.net/ctheory_wp/the-secret-life-of-our-prosthses.

Bamford, James. *The Shadow Factory: The Ultra-Secret NSA from 9/11 to the Eavesdropping on America.* New York: Doubleday, 2008.

Banet-Weiser, Sarah. *Authentic: The Politics of Ambivalence in Brand Culture.* New York: New York University Press, 2012.

Bannister, Frank, and Regina Connolly. "The Trouble with Transparency: A Critical Review of Openness in e-Government." *Policy & Internet* 3, no. 1 (2011): 158–187.

Barker, Greg, dir. *Homegrown: The Counter-Terror Dilemma.* New York: HBO, February 8, 2016.

Barnet, Belinda. "The Secret Life of Our Prostheses." *Ctheory* (2015).

Barocas, Solon, and Karen Levy. "Refractive Surveillance: Monitoring Customers to Manage Workers." *International Journal of Communication* 12 (2007): 1166–1188.

Barrett, Richard. "Don't Turn Security into Theater." CNN, May 6, 2013. http://globalpublicsquare.blogs.cnn.com/2013/05/06/dont-turn-security-into-theater/.

Bateson, Nora. "Warm Data: Contextual Research and New Forms of Information." *Hackernoon*, May 28, 2017. https://hackernoon.com/warm-data-9f0fcd2a828c.

Bauman, Zygmunt, and David Lyon. *Liquid Surveillance: A Conversation*. Cambridge: Polity, 2013.

Bayoumi, Moustafa. "What's a 'Lone Wolf'? It's the Special Name We Give White Terrorists." *The Guardian*, October 2, 2017. https://www.theguardian.com/commentisfree/2017/oct/04/lone-wolf-white-terrorist-las-vegas.

BBC News. "Rolling Stone Defends Boston Bomb Suspect Cover." July 17, 2013. https://www.bbc.com/news/world-us-canada-23351317.

BBC News. "Peeple App for Rating Human Beings Causes Uproar." October 1, 2015. https://www.bbc.com/news/technology-34415382.

BBC News. "Two London Attackers Named by Police." June 5, 2017. https://www.bbc.com/news/uk-40165646.

BBC News. "Las Vegas Shootings: Is the Gunman a Terrorist?" October 3, 2017. https://www.bbc.com/news/world-us-canada-41483943.

BBC News. "John Hancock Adds Fitness Tracking to All Policies." September 20, 2018. https://www.bbc.com/news/technology-45590293.

Beck, Ulrich. *Risk Society: Towards a New Modernity*. London: Sage, 1992.

———. *World at Risk*. Malden, MA: Polity Press, 2009.

Becker, Jo, and Scott Shane. "Secret 'Kill List' Proves a Test of Obama's Principles and Will." *The New York Times*, May 29, 2012. http://www.nytimes.com/2012/05/29/world/obamas-leadership-in-war-on-al-qaeda.html?pagewanted=1&_r=1.

Bell, Genevieve, and Paul Dourish. "Yesterday's Tomorrows: Notes on Ubiquitous Computing's Dominant Vision." *Personal and Ubiquitous Computing* 11, no. 2 (2007): 133–143.

Bennington, Geoffrey. "Kant's Open Secret." *Theory, Culture & Society* 28, no. 7–8 (January 2012): 26–40.

Bergen, Peter L. *United States of Jihad—Investigating America's Homegrown Terrorists*. New York: Crown Publishers, 2016.

Berlant, Lauren. *Cruel Optimism*. Durham, NC: Duke University Press, 2011.

Berry, David M. *The Philosophy of Software: Code and Mediation in the Digital Age*. Basingstoke, UK: Palgrave Macmillan, 2011.

Berson, Josh. *Computable Bodies: Instrumented Life and the Human Somatic Niche*. London: Bloomsbury Academic, 2015.

Betancourt, Michael. "The Demands of Agnotology::Surveillance." *Ctheory*, 2014. http://ctheory.net/ctheory_wp/the-demands-of-agnotologysurveillance.

Beydoun, Khaled. "'Lone Wolf': Our Stunning Double Standard When It Comes to Race and Religion." *The Washington Post*, October 2, 2017. https://www.washingtonpost.com/news/acts-of-faith/wp/2017/10/02/lone-wolf-our-stunning-double-standard-when-it-comes-to-race-and-religion/?utm_term=.1e17de80f30e.

Bhatt, Sarita. "We're All Narcissists Now, and that's a Good Thing." *Fast Company*, September 27, 2013. http://www.fastcoexist.com/3018382/were-all-narcissists-now-and-thats-a-good-thing.

Biddle, Sam. "Amazon's Home Surveillance Chief Declared War on 'Dirtbag Criminals' as Company Got Closer to Police." *The Intercept*, February 14, 2019. https://theintercept.com/2019/02/14/amazon-ring-police-surveillance/.

Birchall, Claire. "Introduction to 'Secrecy and Transparency': The Politics of Opacity and Openness." *Theory, Culture & Society* 28, no. 7–8 (2012): 7–25.

Bjerg, Ole, and Thomas Presskorn-thygesen. "Conspiracy Theory: Truth Claim or Language Game?" *Theory, Culture & Society* 34, no. 1 (2017): 137–159.

Blair, Tony. "Full Text: Tony Blair's Speech." *The Guardian*, March 5, 2004. http://www.theguardian.com/politics/2004/mar/05/iraq.iraq.

Boesel, Whitney Erin. "Return of the Quantrepreneurs." *Society Pages,* Cyborgology section, September 26, 2013. https://thesocietypages.org/cyborgology/2013/09/26/return-of-the-quantrepreneurs/.

Bogost, Ian. "Why Nothing Works Anymore." *The Atlantic*, February 23, 2017. https://www.theatlantic.com/technology/archive/2017/02/the-singularity-in-the-toilet-stall/517551/.

Bohan, Caren. "Lawmakers Urge Review of Domestic Spying, Patriot Act." *Chicago Tribune*, June 9, 2013. http://articles.chicagotribune.com/2013-06-09/news/sns-rt-us-usa-security-lawmakersbre9580ab-20130609_1_guardian-national-security-agency-surveillance.

Borges, Jorge Luis. "The Library of Babel." In *Jorge Luis Borges: Collected Fictions*, translated by Andrew Hurley, 112–118. New York: Penguin Books, 1998.

Borgmann, Albert. "The Moral Significance of the Material Culture." *Inquiry: An Interdisciplinary Journal of Philosophy* 35, no. 3–4 (1992): 291–300.

Borland, John. "Glenn Greenwald: 'A Lot' More NSA Documents to Come." *Wired*, December 27, 2013. http://www.wired.com/2013/12/greenwald-lot-nsa-documents-come/.

Bowden, Mark. "The Measured Man." *The Atlantic* (July/August 2012). http://www.theatlantic.com/magazine/archive/2012/07/the-measured-man/309018/.

boyd, danah, and Kate Crawford. "Critical Questions for Big Data." *Information, Communication & Society* 15, no. 5 (2012): 662–679.

Brandon, John. "6 Tech Trends of the Far Future." *Inc.*, May 16, 2013. http://www.inc.com/john-brandon/6-tech-trends-of-the-far-future.html.

Bratich, Jack. "Public Secrecy and Immanent Security." *Cultural Studies* 20, no. 4–5 (2006): 493–511.

Brayne, Sarah. "Big Data Surveillance: The Case of Policing." *American Sociological Review* 82, no. 5 (2017): 977–1008.

Brouwer, Sam de, Linda Avey, Jessica Richman, and Tan Le. "Frontiers of Tracking Health." Presentation at Quantified Self 2015 Conference, San Francisco, June 18–20, 2015.

Brozen, Yale. "The Value of Technological Change." *Ethics* 62, no. 4 (1952): 249–265.

Brynielsson, Joel, Andreas Horndahl, Fredrik Johansson, Lisa Kaati, Christian Martenson, and Pontus Svenson. "Analysis of Weak Signals for Detecting Lone Wolf Terrorists." In *Intelligence and Security Informatics Conference*, edited by Nasrullah Memon and Daniel Zeng, 197–204. Los Alamitos, CA: IEEE Computer Society, 2012.

Bucher, Taina. "Want to Be on the Top? Algorithmic Power and the Threat of Invisibility on Facebook." *New Media & Society* 14, no. 7 (2012): 1164–1180.

Bunz, Mercedes, and Graham Meikle. *The Internet of Things*. Cambridge: Polity Press, 2018.

Burris, Sarah. "'Minority Report' Is Coming True: We Now Have Threat Scores to Match Our Credit Scores." *Salon*, January 15, 2016. http://www.salon.com/2016/01/15/minority_report_is_coming_true_we_now_have_threat_scores_to_match_our_credit_scores_partner/.

Burton, Fred, and Scott Stewart. "The 'Lone Wolf' Disconnect." *STRATFOR*, January 30, 2008. https://www.stratfor.com/weekly/lone_wolf_disconnect.

Bury, John Bagnell. *The Idea of Progress: An Inquiry into Its Origin and Growth*. London: Macmillan, 1920.

Butler, Judith. *The Psychic Life of Power*. Stanford, CA: Stanford University Press, 1997.

Calamur, Krishnadev. "NYPD Settles Pair of Lawsuits Over Muslim Surveillance." *The Atlantic*, January 7, 2016. http://www.theatlantic.com/national/archive/2016/01/nypd-surveillance-muslims-settlement/423174/.

Calo, Ryan. "Digital Market Manipulation." *George Washington Legal Review* 82, no. 4 (2014): 995–1051.

Carnright, Cassandra B. "Affidavit, United States of America v. Joseph D. Jones and Edward Schimenti," 2017. https://www.justice.gov/usao-ndil/press-release/file/957226/download

Carroll, Rory. "Inspector Gadget: How Smart Devices Are Outsmarting Criminals." *The Guardian*, June 23, 2017. https://www.theguardian.com/technology/2017/jun/23/smart-devices-solve-crime-murder-internet-of-things.

Cassidy, John. "Why Edward Snowden Is a Hero." *The New Yorker*, June 10, 2013. http://www.newyorker.com/news/john-cassidy/why-edward-snowden-is-a-hero.

CBS News. "Dzhokhar and Tamerlan: A Profile of the Tsarnaev Brothers." April 23, 2013. https://www.cbsnews.com/news/dzhokhar-and-tamerlan-a-profile-of-the-tsarnaev-brothers/.

CCS Insight. "Wearables Market to Be Worth $25 Billion by 2019." September 1, 2015. https://www.ccsinsight.com/press/company-news/2332-wearables-market-to-be-worth-25-billion-by-2019-reveals-ccs-insight/.

Ceglowski, Maciej. "Haunted by Data." Presentation at the Strata+Hadoop World Conference, New York, September 29–October 1, 2015.

Chakrabarti, Shami. "Let Me Be Clear—Edward Snowden Is a Hero." *The Guardian*, June 14, 2015. https://www.theguardian.com/commentisfree/2015/jun/14/edward-snowden-hero-government-scare-tactics.

Chamayou, Grégoire. "Oceanic Enemy: A Brief Philosophical History of the NSA." *Radical Philosophy* 191, no. July (2015): 2–12.

Chamorro-Premuzic, Tomas. "Reputation and the Rise of the 'Rating' Society." *The Guardian*, October 26, 2015. http://www.theguardian.com/media-network/2015/oct/26/reputation-rating-society-uber-airbnb.

Chandler, Jennifer A. "'Obligatory Technologies': Explaining Why People Feel Compelled to Use Certain Technologies." *Bulletin of Science, Technology & Society* 32, no. 4 (2012): 255–264.

Cheney-Lippold, John. *We Are Data—Algorithms and the Making of Our Digital Selves*. New York: New York University Press, 2017.

Choe, Eun Kyoung, Nicole B. Lee, Bongshin Lee, Wanda Pratt, and Julie A. Kientz. "Understanding Quantified-Selfers' Practices in Collecting and Exploring Personal Data." In *Proceedings of the 32nd Annual ACM Conference on Human Factors in Computing Systems*, edited by Matt Jones, Philippe Palanque, Albrecht Schmidt, and Tovi Grossman, 1143–1152. Toronto: ACM, 2014.

Chun, Wendy Hui Kyong. "Crisis, Crisis, Crisis, or Sovereignty and Networks." *Theory, Culture & Society* 28, no. 6 (2011): 91–112.

———. *Updating to Remain the Same: Habitual New Media*. Cambridge, MA: MIT Press, 2015.

Citron, Danielle Keats, and Frank Pasquale. "The Scored Society: Due Process for Automated Predictions." *Washington Law Review* 89 (2014): 101–133.

Clarke, Richard A., Michael J. Morell, Geoffrey R. Stone, Cass R. Sunstein, and Peter Swire. "Liberty and Security in a Changing World," 2013.

Clickhole. "We Already Knew The NSA Spies On Us. We Already Know Everything. Everything Is Boring." February 9, 2015. http://www.clickhole.com/article/we-already-knew-nsa-spies-us-we-already-know-every-1876.

Clover. "Clover Provider Manual," 2017.

Coeckelbergh, Mark. "Is Ethics of Robotics about Robots? Philosophy of Robotics beyond Realism and Individualism." *Law, Innovation and Technology* 3, no. 2 (2011): 241–250.

———. *New Romantic Cyborgs: Romanticism, Information Technology, and the End of the Machine*. Cambridge, MA: MIT Press, 2017.

Coffey, Sarah, Patricia Wen, and Matt Carroll. "Bombing Suspect Spent Wednesday as Typical Student." *Boston Globe*, April 19, 2013. https://www.bostonglobe.com/metro/2013/04/19/bombing-suspect-attended-umass-dartmouth-prompting-school-closure-college-friend-shocked-charge-boston-marathon-bomber/8gbczia4qBiWMAPoSQhViO/story.html.

Cohen, Eric. *In the Shadow of Progress: Being Human in the Age of Technology*. New York: New Atlantis Books, 2008.

Cohen, Jon. "The Patient of the Future." *MIT Technology Review*, February 21, 2012. https://www.technologyreview.com/s/426968/the-patient-of-the-future/.

Cohen, Tom. "Military Spy Chief: Have to Assume Russia Knows U.S. Secrets." CNN, March 9, 2014. http://www.cnn.com/2014/03/07/politics/snowden-leaks-russia/.

Cohn, Jonathan Evan. Ultrasonic Bracelet and Receiver for Detecting Position in 2D Plane. US 9881276 B2. USA, issued 2018.

Cole, Matthew, and Robert Windrem. "How Much Did Snowden Take? At Least Three Times Number Reported." NBC News, August 30, 2013. http://www.nbcnews.com/news/other/how-much-did-snowden-take-least-three-times-number-reported-f8C11038702.

Coleman, Gabriella. *Hacker, Hoaxer, Whistleblower, Spy: The Many Faces of Anonymous*. London: Verso, 2014.

Conroy, J. Oliver. "They Hate the US Government, and They're Mul-
    tiplying: The Terrifying Rise of 'Sovereign Citizens.'" *The Guard-
    ian*, May 15, 2017. https://www.theguardian.com/world/2017/may/15/
    sovereign-citizens-rightwing-terrorism-hate-us-government.
Cooper, Melinda. "Pre-Empting Emergence: The Biological Turn in the War on Terror."
    Theory, Culture & Society 23, no. 4 (July 1, 2006): 113–135.
Cryptome. "42 Years for Snowden Docs Release, Free All Now." February 10, 2016.
    http://cryptome.org/2013/11/snowden-tally.htm.
Currier, Cora. "48 Questions the FBI Uses to Determine If Someone Is a Likely Ter-
    rorist." *The Intercept*, February 13, 2017. https://theintercept.com/2017/02/13/48-
    questions-the-fbi-uses-to-determine-if-someone-is-a-likely-terrorist/.
Daase, Christopher, and Oliver Kessler. "Knowns and Unknowns in the 'War on Ter-
    ror': Uncertainty and the Political Construction of Danger." *Security Dialogue* 38,
    no. 4 (2007): 411–434.
Dadich, Scott. "Science Fiction Helps Make Sense of an Uncertain Future." *Wired*,
    December 13, 2016. https://www.wired.com/2016/12/editors-letter-january-2017/.
Daipha, Phaedra. *Masters of Uncertainty: Weather Forecasters and the Quest for Ground
    Truth*. Chicago: Chicago University Press, 2015.
Danter, Stefan, Ulfried Reichardt, and Regina Schober. "Theorising the Quantified Self
    and Posthumanist Agency." *Digital Culture & Society* 2, no. 1 (2016): P53–67.
Daston, Lorraine. "The Moral Economy of Science." *Osiris* 10 (1995): 3–24.
———. "Objectivity and the Escape from Perspective." *Social Studies of Science* 22
    (1992): 597–618.
———, and Peter Galison. *Objectivity*. New York: Zone Books, 2007.
Davidson, Amy. "The N.S.A.'s Spying on Muslim-Americans." *The New
    Yorker*, July 10, 2014. http://www.newyorker.com/news/amy-davidson/
    the-n-s-a-s-spying-on-muslim-americans.
———. "Unclear Dangers." The *New Yorker*, May 18, 2015. http://www.newyorker.com/
    magazine/2015/05/18/unclear-dangers.
Dean, Jodi. "Publicity's Secret." *Political Theory* 29, no. 5 (2001): 624–650.
Dean, Mitchell. "Risk, Calculable and Incalculable." *Soziale Welt* 49 (1998): 25–42.
de Beauvoir, Simone. *The Ethics of Ambiguity*. New York: Philosophical Library, 1948.
Delany, Samuel R. "About Five Thousand One Hundred and Seventy Five Words." In *Sf:
    The Other Side of Realism*, edited by Thomas D. Clareson, 130–146. Bowling Green,
    OH: Bowling Green University Popular Press, 1971.
Deleuze, Gilles. "Postscript on the Societies of Control." *October* 59 (1992): 3–7.
Den Hond, Chris, and Angela Charlton. "French Candidate Macron Wants to
    Fix Suburban Unrest." Associated Press, March 8, 2017. https://apnews.com/
    fbc665e7cde842db95f8382abd0d81de.
Derrida, Jacques. *Archive Fever: A Freudian Impression*. Chicago: Chicago University
    Press, 1998.
———, and Maurizio Ferraris. *A Taste for the Secret*. Edited by Giacomo Donis and
    David Webb. Translated by Giacomo Donis. Cambridge: Polity Press, 2001.

Devereaux, Ryan. "Manhunting in the Hindu Kush." *The Intercept*, October 15, 2015. https://theintercept.com/drone-papers/manhunting-in-the-hindu-kush/.

Dewey, Caitlin. "Everyone You Know Will Be Able to Rate You on the Terrifying 'Yelp for People'—Whether You Want Them to or Not." *The Washington Post*, September 30, 2015. https://www.washingtonpost.com/news/the-intersect/wp/2015/09/30/everyone-you-know-will-be-able-to-rate-you-on-the-terrifying-yelp-for-people-whether-you-want-them-to-or-not/.

Doctorow, Cory. "Exclusive: Snowden Intelligence Docs Reveal UK Spooks' Malware Checklist." *Boingboing*, February 2, 2016. http://boingboing.net/2016/02/02/doxxing-sherlock-3.html.

Dohrn-van Rossum, Gerhard. *History of the Hour: Clocks and Modern Temporal Orders*. Chicago: University of Chicago Press, 1996.

Douglas, Mary. "Dealing with Uncertainty." *Ethical Perspectives* 8, no. 3 (2001): 145–155.

Douglas, Thomas. "Human Enhancement and Supra-Personal Moral Status." *Philosophical Studies* 162, no. 3 (2013): 473–497.

Doyle, Arthur Conan. *The Sign of the Four*. Stilwell, KS: Digireads.com, 2005.

Dumit, Joseph. *Drugs for Life: How Pharmaceutical Companies Define Our Health*. Durham, NC: Duke University Press, 2012.

Eby, Charles A. "The Nation that Cried Lone Wolf: A Data-Driven Analysis of Individual Terrorists in the United States Since 9/11." Thesis, Naval Postgraduate School, Monterey, CA, 2012.

Eck, John E., and Edward R. Maguire. "Have Changes in Policing Reduced Violent Crime? An Assessment of the Evidence." In *The Crime Drop in America*, edited by Alfred Blumstein and Joel Wallman, 207–265. Cambridge: Cambridge University Press, 2000.

Edgerton, David. "Innovation, Technology, or History: What Is the Historiography of Technology about?" *Technology and Culture* 51, no. 3 (2010): 680–697.

"Edward Snowden SXSW: Full Transcript and Video." *Inside. com*, March 10, 2014. http://blog.inside.com/blog/2014/3/10/edward-snowden-sxsw-full-transcription-and-video.

Eisenberg, Eric M. "Ambiguity as Strategy in Organisational Communication." *Communication Monographs* 51, no. 3 (1984): 227–242.

Eliasoph, Nina. *Avoiding Politics: How Americans Produce Apathy in Everyday*. Cambridge: Cambridge University Press, 1998.

Elliott, Justin. "NSA 54 Events Chart." DocumentCloud, 2013. https://www.documentcloud.org/documents/802269-untitled0001.html.

———, and Theodoric Meyer. "Claim on 'Attacks Thwarted' by NSA Spreads Despite Lack of Evidence." *ProPublica*, October 23, 2013. http://www.propublica.org/article/claim-on-attacks-thwarted-by-nsa-spreads-despite-lack-of-evidence.

Ellul, Jacques. *The Technological Society*. New York: Knopf, 1964.

Engelhardt, Tom. *Shadow Government: Surveillance, Secret Wars, and a Global Security State in a Single-Superpower World*. Chicago: Haymarket Books, 2014.

Erickson, Paul, Judy L. Klein, Lorraine Daston, Rebecca Lemov, Thomas Sturm, and Michael D. Gordin. *How Reason Almost Lost Its Mind*. Chicago: University of Chicago Press, 2013.

Erlanger, Steven. "After London Attack, Prime Minister Says, 'Enough Is Enough.'" *The New York Times*, June 4, 2017. https://www.nytimes.com/2017/06/04/world/europe/uk-london-attacks.html?_r=0.

Evans, Robert. "Shitposting, Inspirational Terrorism, and the Christchurch Mosque Massacre." *bellingcat*, March 15, 2019. https://www.bellingcat.com/news/rest-of-world/2019/03/15/shitposting-inspirational-terrorism-and-the-christchurch-mosque-massacre/.

Fallows, James. "Truth, Lies, Politics, and the Press, in Three Acts." *The Atlantic*, August 18, 2012. https://www.theatlantic.com/politics/archive/2012/08/truth-lies-politics-and-the-press-in-three-acts/261297/.

Fantz, Ashley. "NSA Leaker Ignites Global Debate: Hero or Traitor?" CNN, October 6, 2013. http://www.cnn.com/2013/06/10/us/snowden-leaker-reaction/.

Farivar, Cyrus. "Snowden Distributed Encrypted Copies of NSA Files across the World." *Wired*, January 6, 2013. http://www.wired.co.uk/news/archive/2013-06/26/edward-snowden-nsa-data-copies.

Farocki, Harun. "Phantom Images." *Public* 29 (2004): 12–22.

Fenn, Jackie, and Hung LeHong. "Hype Cycle for Emerging Technologies, 2011." Gartner, July 28, 2011. https://www.gartner.com/doc/1754719/hype-cycle-emerging-technologies-.

Fick, Nate. "Was Snowden Hero or Traitor? Perhaps a Little of Both." The *Washington Post*, January 19, 2017. https://www.washingtonpost.com/opinions/was-snowden-hero-or-traitor-perhaps-a-little-of-both/2017/01/19/a2b8592e-c6f0-11e6-bf4b-2c064d32a4bf_story.html?utm_term=.28a8e28234ab.

Fields, Gary, and Evan Perez. "FBI Seeks to Target Lone Extremists." *The Wall Street Journal*, June 15, 2009. http://www.wsj.com/articles/SB124501849215613523.

Finley, Klint. "Interview: Sensor Hacking for Mindfulness with Nancy Dougherty on the New Mindful Cyborgs." *Technoccult*, June 10, 2013. http://technoccult.net/archives/2013/06/10/interview-sensor-hacking-for-mindfulness-with-nancy-dougherty-on-the-new-mindful-cyborgs/.

Finn, Ed. *What Algorithms Want: Imagination in the Age of Computing*. Cambridge, MA: MIT Press, 2017.

Fisher, Mark. *Capitalist Realism: Is There No Alternative?* Winchester, UK: O Books, 2009.

Fleitz, Fred. "Snowden Is a Traitor and a Fraud, Period." *National Review*, September 16, 2016. http://www.nationalreview.com/article/440113/edward-snowden-report-house-intelligence-committee-confirms-he-shouldnt-be-pardoned.

Floridi, Luciano. "Robots, Jobs, Taxes, and Responsibilities." *Philosophy and Technology* 30, no. 1 (2017): 1–4.

Flyverbom, Mikkel. "Transparency: Mediation and the Management of Visibilities." *International Journal of Communication* 10, no. 1 (2016): 110–122.

Foss, Adam, Samuel Sinyangwe, Julia Angwin, Charmaine Arthur, and Kim Foxx. "Automating (In)Justice: Policing and Sentencing in the Algorithm Age." Presentation at Data 4 Black Lives, Cambridge, MA, October 17–19, 2017.

Foster, John Bellamy, and Robert W. McChesney. "Surveillance Capitalism: Monopoly-Finance Capital, the Military-Industrial Complex, and the Digital Age." *Monthly Review* 66, no. 3 (2014). https://monthlyreview.org/2014/07/01/surveillance-capitalism/.

Fotopoulou, Aristea, and Kate O'Riordan. "Biosensory Experiences and Media Materiality Fitbit and Biosensors: Imaginaries and Material Instantiations." Presentation at IR16, Phoenix, AZ, October 21–24, 2015.

Foucault, Michel. *The Order of Things: An Archaeology of the Human Sciences.* London: Routledge, 2002.

———. *Security, Territory, Population: Lectures at the Collège de France 1977–1978.* Edited by Michel Senellart. Translated by Graham Burchell. New York: Palgrave Macmillan, 2004.

———. *The Birth of Biopolitics: Lectures at the Collége de France, 1978–79.* Edited by Michel Senellart. Translated by Graham Burchell. Basingstoke, UK: Palgrave Macmillan, 2008.

———. *The Courage of Truth: Lectures at the Collège de France, 1983–1984.* Edited by Frédéric Gros. Translated by Graham Burchell. Basingstoke, UK: Palgrave Macmillan, 2011.

———. *Wrong-Doing, Truth-Telling: The Function of Avowal in Justice.* Edited by Fabienne Brion and Bernard E. Harcourt. Chicago: University of Chicago Press, 2014.

———. *On the Government of the Living: Lectures at the Collège de France, 1979–1980.* Basingstoke, UK: Palgrave Macmillan, 2014.

———. *The Punitive Society: Lectures at the College de France 1972–1973.* Edited by Bernard E. Harcourt. New York: Palgrave Macmillan, 2015.

———. *Speech Begins after Death.* Minneapolis: University of Minnesota Press, 2017.

Fox News. "Edward Snowden: Whistleblower or Double Agent?" June 14, 2013. https://www.foxnews.com/politics/edward-snowden-whistleblower-or-double-agent.

Franklin, Benjamin. *The Autobiography of Benjamin Franklin.* London: George Bell & Sons, 1884.

Freedman, Carl. "Towards a Theory of Paranoia: The Science Fiction of Philip K. Dick." *Science Fiction Studies* 11, no. 1 (1984): 15–24.

Frers, Lars. "The Matter of Absence." *Cultural Geographies* 20, no. 4 (2013): 431–445.

Freund, Peter E. S. "Civilised Bodies Redux: Seams in the Cyborg." *Social Theory & Health* 2, no. 3 (2004): 273–289.

Frické, Martin. "The Knowledge Pyramid: A Critique of the DIKW Hierarchy." *Journal of Information Science* 35, no. 2 (2009): 131–142.

Friedersdorf, Conor. "New Surveillance Whistleblower: The NSA Violates the Constitution." *The Atlantic,* July 21, 2014. http://www.theatlantic.com/politics/archive/2014/07/a-new-surveillance-whistleblower-emerges/374722/.

Frog. "15 Tech Trends that Will Define 2014, Selected By Frog." *Fast Company*, January 8, 2014. http://www.fastcodesign.com/3024464/15-tech-trends-that-will-define-2014-selected-by-frog.

Fuchs, Wolfgang Walter. *Phenomenology and the Metaphysics of Presence*. The Hague: Martinus Nijhoff, 1976.

Funtowicz, Silvio O., and Jerome R. Ravetz. "The Emergence of Post-Normal Science." In *Science, Politics and Morality: Scientific Uncertainty and Decision Making*, edited by Rene von Schomberg, 85–123. Berlin: Springer-Science+Business Media, B.V., 1993.

Gallagher, Ryan. "Latest Documents From Snowden Provide Direct Proof of Unlawful Spying on Americans." *Slate*, August 16, 2013, http://www.slate.com/blogs/future_tense/2013/08/16/latest_snowden_documents_prove_proof_of_unlawful_spying_on_americans.html.

———, and Henrik Moltke. "Titanpointe: The NSA's Spy Hub in New York, Hidden in Plain Sight." *The Intercept*, November 16, 2016. https://theintercept.com/2016/11/16/the-nsas-spy-hub-in-new-york-hidden-in-plain-sight/.

Gale, Catherine, prod. and dir. *The Joy of Data*. London, UK: BBC Four, 2016.

Gan, Vicky. "How TV's 'Person of Interest' Helps Us Understand the Surveillance Society." *Smithsonian.com*, October 24, 2013. http://www.smithsonianmag.com/smithsonian-institution/how-tvs-person-of-interest-helps-us-understand-the-surveillance-society-5407171/?no-ist.

Gandy, Oscar. *The Panoptic Sort: A Political Economy of Personal Information*. Boulder, CO: Westview Press, 1993.

Garber, Megan. "The Ennui of the Fitbit." *The Atlantic*, July 10, 2015. https://www.theatlantic.com/technology/archive/2015/07/the-ennui-of-the-fitbit/398129/.

Garten, Ariel. "Know Thyself, with a Brain Scanner." Filmed at TEDxToronto, Toronto, Canada, September 2011. Video, 14:49. https://www.ted.com/talks/ariel_garten_know_thyself_with_a_brain_scanner/transcript?language=en.

———, and Mikey Siegel. "Brainwave Technology and Consciousness Hacking." Presentation at the Quantified Self 2015 Conference, San Francisco, June 18–20, 2015.

Gartenstein-Ross, Daveed, and Nathaniel Barr. "The Myth of Lone-Wolf Terrorism." *Foreign Affairs*, July 26, 2016. https://www.foreignaffairs.com/articles/western-europe/2016-07-26/myth-lone-wolf-terrorism.

Gates, Kelly A. *Our Biometric Future: Facial Recognition Technology and the Culture of Surveillance*. New York: New York University Press, 2011.

Gawne, Mark. "The Modulation and Ordering of Affect: From Emotion Recognition Technology to the Critique of Class Composition." *Fibreculture* 21 (2012): 98–123.

Geels, Frank W., and Wim A. Smit. "Lessons from Failed Technological Futures: Potholes in the Road to the Future." In *Contested Futures: A Sociology of Prospective Techno-Science*, edited by Nik Brown, Brian Rappert, and Andrew Webster, 129–155. Aldershot, UK: Ashgate, 2000.

Gekker, Alex. "Casual Power." *Digital Culture & Society* 2, no. 1 (2016): 107–122.

Gellman, Barton, and Matt DeLong. "The NSA's Problem? Too Much Data." *The Washington Post*, October 14, 2013. http://apps.washingtonpost.com/g/page/world/the-nsas-overcollection-problem/517/.

Gerecht, Reuel Marc. "The Costs and Benefits of the NSA." *The Weekly Standard*, June 24, 2013. http://www.weeklystandard.com/article/costs-and-benefits-nsa/735246.

Gerstein, Josh. "Spies Prep Reporters on Protecting Secrets." *New York Sun*, September 27, 2007. http://www.nysun.com/national/spies-prep-reporters-on-protecting-secrets/63465/.

Gill, Paul, John Horgan, and Paige Deckert. "Bombing Alone: Tracing the Motivations and Antecedent Behaviors of Lone-Actor Terrorists." *Journal of Forensic Sciences* 59, no. 2 (March 2014): 425–435.

Gillespie, Tarleton. "The Politics of 'Platforms.'" *New Media & Society* 12, no. 3 (2010): 347–364.

———. "#Trendingistrending: When Algorithms Become Culture." In *Algorithmic Cultures: Essays on Meaning, Performance and New Technologies,*, edited by Robert Seyfert and Jonathan Roberge, 52–75. Abingdon, UK: Routledge, 2016.

Ginzburg, Carlo. "Morelli, Freud and Sherlock Holmes: Clues and Scientific Method." *History Workshop Journal* 9 (1980): 5–36.

Gitelman, Lisa, and Virginia Jackson. "Introduction." In *Raw Data Is an Oxymoron*, edited by Lisa Gitelman, 1–14. Cambridge, MA: MIT Press, 2013.

Goetz, Thomas. "The Diabetic's Paradox." *The Atlantic*, April 1, 2013. http://www.theatlantic.com/health/archive/2013/04/the-diabetics-paradox/274507/.

Goffman, Alice. *On the Run: Fugitive Life in an American City*. Chicago: University of Chicago Press, 2014.

Goldberg, Jeffrey. "The Obama Doctrine." *The Atlantic*, March 10, 2016. http://www.theatlantic.com/magazine/archive/2016/04/the-obama-doctrine/471525/#article-comments.

Goldman, Adam. "The NSA Has No Idea How Much Data Edward Snowden Took because He Covered His Digital Tracks." *Business Insider*, August 24, 2013. http://www.businessinsider.com/edward-snowden-covered-tracks-2013-8.

Golumbia, David. *The Cultural Logic of Computation*. Cambridge, MA: Harvard University Press, 2009.

Gould, Stephen Jay. *The Mismeasure of Man: The Definitive Refutation to the Argument of The Bell Curve*. New York: W. W. Norton, 1996.

Granick, Jennifer Stisa. *American Spies: Modern Surveillance, Why You Should Care, and What to Do about It*. Cambridge: Cambridge University Press, 2017.

Gray, Mary L., and Siddharth Suri. *Ghost Work: How to Stop Silicon Valley from Building a New Global Underclass*. Boston: Houghton Mifflin Harcourt, 2019.

Greenberg, Andy. "How to Anonymize Everything You Do Online." *Wired*, June 17, 2014. http://www.wired.com/2014/06/be-anonymous-online/.

Greenfield, Dana. "Deep Data: Notes on the n of 1." In *Quantified—Biosensing Technologies in Everyday Life*, edited by Dawn Nafus, 123–146. Cambridge, MA: MIT Press, 2016.

Greenwald, Glenn. "'Explosive' NSA Spying Reports Are Imminent." *Der Spiegel*, July 19, 2013. http://www.spiegel.de/international/world/journalist-says-explosive-reports-coming-from-snowden-data-a-912034.html.

———. *No Place to Hide: Edward Snowden, the NSA and the Surveillance State*. London: Penguin Books, 2014.

Gregg, Melissa. *Counterproductive: Time Management in the Knowledge Economy*. Durham, NC: Duke University Press, 2018.

———. *Work's Intimacy*. New York: Polity, 2011.

Gregory, Derek. "From a View to a Kill: Drones and Late Modern War." *Theory, Culture & Society* 28, no. 7–8 (January 12, 2011): 188–215.

Griffiths, Sarah. "Is This the Most Connected Human on the Planet? Man Is Wired up to 700 Sensors to Capture Every Single Detail of His Existence." *Daily Mail*, March 25, 2014. http://www.dailymail.co.uk/sciencetech/article-2588779/Is-connected-man-planet-Man-wired-700-devices-capture-single-existence.html.

Grunwald, Armin. "Against Over-Estimating the Role of Ethics in Technology Development." *Science and Engineering Ethics* 6 (2000): 181–196.

Gumbrecht, Hans Ulrich. *Production of Presence: What Meaning Cannot Convey*. Stanford, CA: Stanford University Press, 2003.

Gunkel, David J. *The Machine Question: Critical Perspectives on AI, Robots, and Ethics*. Cambridge, MA: MIT Press, 2012.

Hacking, Ian. "Statistical Language, Statistical Truth and Statistical Reason: The Self-Authentification of a Style of Scientific Reasoning." In *Social Dimensions of Science*, edited by Ernan McMullin, 130–157. Notre Dame, IN: University of Notre Dame Press, 1992.

———. *The Taming of Chance*. Cambridge: Cambridge University Press, 1990.

Hadjimatheou, Katerina. "Surveillance Technologies, Wrongful Criminalisation, and the Presumption of Innocence." *Philosophy and Technology* 30, no. 1 (2017): 39–54.

Haggerty, Kevin D., and Richard V. Ericson. "The Surveillant Assemblage." *British Journal of Sociology* 51, no. 4 (2000): 605–622.

Halpern, Orit. *Beautiful Data: A History of Vision and Reason since 1945*. Durham, NC: Duke University Press, 2014.

Hannah, Matthew G. "(Mis)Adventures in Rumsfeld Space." *GeoJournal* 75, no. 4 (2010): 397–406.

Hansen, Mark B. N. "Engineering Pre-Individual Potentiality: Technics, Transindividuation, and 21st-Century Media." *SubStance* 41, no. 3 (2012): 32–59.

———. *Feed-Forward: On the Future of Twenty-First-Century Media*. Chicago: University of Chicago Press, 2015.

Hansson, Sven Ove. "From the Casino to the Jungle: Dealing with Uncertainty in Technological Risk Management." *Synthese* 168, no. 3 (2009): 423–432.

———. "Philosophical Perspectives on Risk." Presentation at Research in Ethics and Engineering, Delft, Netherlands, 2002.

Harcourt, Bernard. *Against Prediction—Profiling, Policing, and Punishing in an Actuarial Age*. Chicago: Chicago University Press, 2007.

———. *Exposed: Desire and Disobedience in the Digital Age.* Cambridge, MA: Harvard University Press, 2015.

Harrington, Anne. *The Cure Within: A History of Mind-Body Medicine.* New York: W. W. Norton, 2008.

Harris, Shane. *The Watchers: The Rise of America's Surveillance State.* New York: Penguin Press, 2010.

Harvey, David. *Marx, Capital, and the Madness of Economic Reason.* New York: Oxford University Press, 2018.

———. *The New Imperialism.* Oxford: Oxford University Press, 2003.

Haskin, David. "Don't Believe the Hype: The 21 Biggest Technology Flops." *Computerworld,* April 4, 2007. https://www.computerworld.com/article/2543763/computer-hardware/don-t-believe-the-hype--the-21-biggest-technology-flops.html.

Hassein, Nabil. "Against Black Inclusion in Facial Recognition." *Digital Talking Drum,* August 15, 2017. https://digitaltalkingdrum.com/2017/08/15/against-black-inclusion-in-facial-recognition/.

Havens, John C. *Hacking H(App)Iness: Why Your Personal Data Counts and How Tracking It Can Change the World.* New York: Jeremy P. Tarcher/Penguin, 2014.

Hayden, Michael. *Playing to the Edge: American Intelligence in the Age of Terror.* New York: Penguin Press, 2016.

———. Address to the National Press Club, Washington, DC, January 23, 2006.

———. Statement for the Record to the Joint Inquiry of the Senate Select Committee on Intelligence and the House Permanent Select Committee on Intelligence. October 17, 2002.

Hayles, N. Katherine. *How We Became Posthuman: Virtual Bodies in Cybernetics, Literature, and Informatics.* Chicago: University of Chicago Press, 1999.

———. *My Mother Was A Computer: Digital Subjects and Literary Texts.* Chicago: University of Chicago Press, 2005.

———. "Cognitive Assemblages: Technical Agency and Human Interactions." *Critical Inquiry* 43, no. 1 (2016): 32–55.

———. "The Cognitive Nonconscious: Enlarging the Mind of the Humanities." *Critical Inquiry* 42 (2016): 783–808.

———. *Unthought: The Power of the Cognitive Nonconscious.* Chicago: University of Chicago Press, 2017.

Heidegger, Martin. "The Question Concerning Technology," 1949. http://www.wright.edu/cola/Dept/PHL/Class/P.Internet/PITexts/QCT.html.

Hello.is. "Meet Sense." Accessed April 13, 2016. https://hello.is/videos#meet-sense.

Henderson, Barney. "Boston Marathon Bombs: Suspect Captured—April 20 as It Happened." *The Telegraph,* April 20, 2013. http://www.telegraph.co.uk/news/worldnews/northamerica/usa/10007370/Boston-Marathon-bombs-suspect-captured-April-20-as-it-happened.html.

Hendrix, Jenny. "NSA Surveillance Puts George Orwell's '1984' on Bestseller Lists." *Los Angeles Times,* June 11, 2013. http://articles.latimes.com/2013/jun/11/entertainment/la-et-jc-nsa-surveillance-puts-george-orwells-1984-on-bestseller-lists-20130611.

Henley, Jon. "French Presidential Candidates Debate Key Election Issues—as It Happened." *The Guardian*, March 20, 2017. https://www.theguardian.com/world/live/2017/mar/20/french-presidential-candidates-debate-key-election-issues-live.

Hernandez, Daniela. "Big Data Is Transforming Healthcare." *Wired*, October 16, 2012. http://www.wired.com/2012/10/big-data-is-transforming-healthcare/.

Hesse, Monica. "Bytes of Life." The *Washington Post*, September 9, 2008. http://www.wired.com/2010/11/mf_qa_ferriss/.

Hill, Eleanor. Joint Inquiry Staff Statement, Hearing on the Intelligence Community's Response to Past Terrorist Attacks Against the United States from February 1993 to September 2001, 2002. http://fas.org/irp/congress/2002_hr/100802hill.html.

Hofstadter, Richard. "The Paranoid Style in American Politics." In *The Paranoid Style in American Politics and Other Essays*, 3–40. New York: Vintage Books, 1967.

Holland, Emma. "Finally, a Better Way for 'Quantified Self' Products to Collect Personal Data." *Fast Company*, October 18, 2013. http://www.fastcompany.com/3020212/finally-a-better-way-for-quantified-self-products-to-collect-personal-data.

Hong, Sun-ha. "The Other-Publics: Mediated Othering and the Public Sphere in the Dreyfus Affair." *European Journal of Cultural Studies* 17, no. 6 (2014): 665–681.

Hörl, Erich. "The Technological Condition." *Parrhesia* 22 (2015): 1–15.

Horn, Eva. "Logics of Political Secrecy." *Theory, Culture & Society* 28, no. 7–8 (2012): 103–122.

Horning, Rob. "Google Alert for the Soul." *The New Inquiry*, April 12, 2013. http://thenewinquiry.com/essays/google-alert-for-the-soul/.

Hosenball, Mark. "NSA Chief Says Snowden Leaked up to 200,000 Secret Documents." Reuters, November 14, 2013. http://www.reuters.com/article/us-usa-security-nsa-idUSBRE9AD19B20131114.

Hu, Tung-Hui. *A Prehistory of the Cloud*. Cambridge, MA: MIT Press, 2015.

Human Rights Watch and ACLU. *With Liberty to Monitor All: How Large-Scale US Surveillance Is Harming Journalism, Law, and American Democracy*. New York: Human Rights Watch, 2014.

"If It Weren't for Edward Snowden Conspiracy Theories Would Still Just Be 'Theories.'" Reddit, March 15, 2015. https://www.reddit.com/r/conspiracy/comments/2z31bh/if_it_werent_for_edward_snowden_conspiracy/.

Igo, Sarah E. *The Known Citizen: A History of Privacy in Modern America*. Cambridge, MA: Harvard University Press, 2018.

Illouz, Eva. *Cold Intimacies: The Making of Emotional Capitalism*. Malden, MA: Polity Press, 2007.

———. *Saving the Modern Soul: Therapy, Emotions, and the Culture of Self-Help*. Vol. 38. Berkeley: University of California Press, 2008.

Human Rights Watch. "Illusion of Justice: Human Rights Abuses in US Terrorism Prosecutions," 2014. https://www.hrw.org/sites/default/files/reports/usterrorism0825_ForUpload_1_0.pdf.

International Data Corporation. "Worldwide Wearables Market Soars in the Third Quarter as Chinese Vendors Challenge the Market Leaders, According to IDC." December 3, 2015. http://www.idc.com/getdoc.jsp?containerId=prUS40674715.

Israel, Jonathan I. *Radical Enlightenment: Philosophy and the Making of Modernity 1650–1750*. Oxford: Oxford University Press, 2001.

Jakicic, John M., Kelliann K. Davis, and Renee J. Rogers. "Effect of Wearable Technology Combined with a Lifestyle Intervention on Long-Term Weight Loss." *JAMA: The Journal of the American Medicine Association* 316, no. 11 (2016): 1161–1171.

Jasanoff, Sheila. *Science at the Bar: Law, Science, and Technology in America*. Cambridge: Harvard University Press, 1995.

———. "Future Imperfect: Science, Technology, and the Imaginations of Modernity." In *Dreamscapes of Modernity: Sociotechnical Imaginaries and the Fabrication of Power*, edited by Sheila Jasanoff and Sang-Hyun Kim, 1–33. Chicago: Chicago University Press, 2015.

———. *The Ethics of Invention: Technology and the Human Future*. New York: W. W. Norton, 2016.

Jenkins Jr., Holman W. "Google and the Search for the Future." *The Wall Street Journal*, August 14, 2010. http://www.wsj.com/articles/SB10001424052748704901104575423294099527212.

Joh, Elizabeth E. "Feeding the Machine: Policing, Crime Data, & Algorithms." *William & Mary Bill of Rights Journal* 26, no. 3 (2017): 287–306.

Johnson, Alex. "Edward Snowden 'Probably' Not a Russian Spy, New NSA Chief Says." NBC News, June 3, 2014. http://www.nbcnews.com/storyline/nsa-snooping/edward-snowden-probably-not-russian-spy-new-nsa-chief-says-n121926.

Jonas, Steven. "What We Are Reading." *Quantified Self*, October 26, 2015. http://quantifiedself.com/2015/10/reading-73/.

Jordan, Matthew, and Nikki Pfarr. "Forget the Quantified Self. We Need to Build the Quantified Us." *Wired*, April 4, 2014. http://www.wired.com/2014/04/forget-the-quantified-self-we-need-to-build-the-quantified-us/.

Jouvenal, Justin. "The New Way Police Are Surveilling You: Calculating Your Threat 'Score.'" The *Washington Post*, January 10, 2016. https://www.washingtonpost.com/local/public-safety/the-new-way-police-are-surveilling-you-calculating-your-threat-score/2016/01/10/e42bccac-8e15-11e5-baf4-bdf37355da0c_story.html?utm_term=.5b5f41b1fe23.

Kaati, Lisa, and Pontus Svenson. "Analysis of Competing Hypothesis for Investigating Lone Wolf Terrorists." In *European Intelligence and Security Informatics Conference*, 295–299. Washington, DC: IEEE Computer Society, 2011.

Kaczynski, Andrew. "Former NSA Director On Edward Snowden: 'He's Working For Someone.'" *BuzzFeed*, June 3, 2014. http://www.buzzfeed.com/andrewkaczynski/former-nsa-director-on-edward-snowden-hes-working-for-someon#.pf97er37p.

Kant, Immanuel. "An Answer to the Question: What Is Enlightenment?" In *What Is Enlightenment? Eighteenth-Century Answers and Twentieth-Century Questions*, edited by James Schmidt, translated by James Schimdt, 58–77. Berkeley: University of California Press, 1996.

Kaye, Kate. "FTC: Fitness Apps Can Help You Shred Calories—and Privacy." *AdAge*, May 7, 2014. http://adage.com/article/privacy-and-regulation/ftc-signals-focus-health-fitness-data-privacy/293080/.

Kelley, Michael B. "The Guardian's Bombshell Revelation About NSA Domestic Spying Is Only the Tip of the Iceberg." *Business Insider*, June 6, 2013. https://www.businessinsider.com/the-impact-of-nsa-domestic-spying-2013-6.

———. "Snowden Has One Very Important and Potentially Devastating Question to Answer." *Business Insider*, March 19, 2014. http://www.businessinsider.com/snowden-and-military-information-2014-3.

Kelly, Kevin. "What Is the Quantified Self?" *Quantified Self*, October 5, 2007. https://web.archive.org/web/20150408202734/http://quantifiedself.com/2007/10/what-is-the-quantifiable-self/.

———. "Self-Tracking? You Will." KK, March 25, 2011. http://kk.org.

Kelly, Samantha Murphy. "The Most Connected Man Is You, Just a Few Years From Now." *Mashable*, January 8, 2014. http://mashable.com/2014/08/21/most-connected-man/#I6oSjAremkqw.

Kessler, Sarah. "Can the Quantified Self Go Too Far?" *Fast Company*, August 19, 2013. http://www.fastcompany.com/3015762/bed-bath-and-beyond-where-do-you-draw-the-line-when-it-comes-to-self-quantifying.

Khosla, Vinod. "The Algorithms Are Coming. What's at Stake?" Presentation at the Quantified Self 2015 Conference, San Francisco, June 18–20, 2015.

King, Shaun. "The White Privilege of the 'Lone Wolf' Shooter." *The Intercept*, October 2, 2017. https://theintercept.com/2017/10/02/lone-wolf-white-privlege-las-vegas-stephen-paddock/.

Kirn, Walter. "If You're Not Paranoid, You're Crazy." The *Atlantic*, November 1, 2015. http://www.theatlantic.com/magazine/archive/2015/11/if-youre-not-paranoid-youre-crazy/407833/.

Kitchin, Rob. "Big Data, New Epistemologies and Paradigm Shifts." *Big Data & Society* 1, no. 1 (2014). https://www.theoryculturesociety.org/kittler-on-the-nsa/.

———, and Gavin McArdle. "What Makes Big Data, Big Data? Exploring the Ontological Characteristics of 26 Datasets." *Big Data & Society* 3, no. 1 (2016): 1–10.

Kittler, Friedrich A. *Gramophone, Film, Typewriter*. Translated by Geoffrey Winthrop-Young and Michael Wutz. Stanford, CA: Stanford University Press, 1986.

———. "No Such Agency." *Theory, Culture & Society*, 2014.

Kline, Ron. "Technological Determinism." In *International Encyclopedia of the Social and Behavioral Sciences*, edited by Neil J. Smelser and Paul B. Baltes, 15495–15498. New York: Elsevier, 2001.

Kluitenberg, Eric. "On the Archaeology of Imaginary Media." In *Media Archaeology: Approaches, Applications, and Implications*, edited by Erkki Huhtamo and Jussi Parikka, 48–69. Berkeley: University of California Press, 2011.

Knappenberger, Brian. "Why Care about the N.S.A.?" *The New York Times*, November 26, 2013. http://www.nytimes.com/video/opinion/100000002571435/why-care-about-the-nsa.html.

Knight, Frank H. *Risk, Uncertainty and Profit*. New York: Augustus M. Kelley, 1964.

Knorr Cetina, Karin. *Epistemic Cultures—How the Sciences Make Knowledge*. Cambridge, MA: Harvard University Press, 1999.

Knowlton, Brian. "Feinstein 'Open' to Hearings on Surveillance Programs." *The New York Times*, June 9, 2013. http://thecaucus.blogs.nytimes.com/2013/06/09/lawmaker-calls-for-renewed-debate-over-patriot-act/?_php=true&_type=blogs&_r=0.

Koopman, Colin. *How We Became Our Data: A Genealogy of the Information Person.* Chicago: Chicago University Press, 2019.

Kozaryn, Linda D. "Alleged Al Qaeda 'Dirty Bomb' Operative in U.S. Military Custody." US Department of Defense, June 10, 2002. http://archive.defense.gov/news/newsarticle.aspx?id=43767.

Krauss, Lawrence M. "Thinking Rationally about Terror." *The New Yorker*, January 2, 2016. http://www.newyorker.com/news/news-desk/thinking-rationally-about-terror?intcid=mod-most-popular.

Kwan, Mei-po. "Algorithmic Geographies: Big Data, Algorithmic Uncertainty, and the Production of Geographic Knowledge." *Annals of the American Association of Geographers* 106, no. 2 (2016): 274–282.

Lake, Eli. "Spy Chief: We Should've Told You We Track Your Calls." *The Daily Beast*, February 17, 2014. http://www.thedailybeast.com/articles/2014/02/17/spy-chief-we-should-ve-told-you-we-track-your-calls.html.

Lambert, Alex. "Bodies, Mood and Excess." *Digital Culture & Society* 2, no. 1 (2016): 71–88.

Lamothe, Dan. "Why Operation Jade Helm 15 Is Freaking out the Internet— and Why It Shouldn't Be." *The Washington Post*, March 31, 2015. https://www.washingtonpost.com/news/checkpoint/wp/2015/03/31/why-the-new-special-operations-exercise-freaking-out-the-internet-is-no-big-deal/.

Landes, David. *The Unbound Prometheus: Technological Change and Industrial Development in Western Europe from 1750 to the Present.* Cambridge: Cambridge University Press, 1969.

Larocca, Amy. "The Wellness Epidemic." *Thecut*, June 27, 2017. https://www.thecut.com/2017/06/how-wellness-became-an-epidemic.html.

Laskow, Sarah. "A New Film Shows How Much We Knew, Pre-Snowden, about Internet Surveillance." *Columbia Journalism Review*, July 15, 2013. http://www.cjr.org/cloud_control/a_new_film_shows_exactly_how_m.php.

Latour, Bruno. "On Technical Mediation—Philosophy, Sociology, Genealogy." *Common Knowledge* 3, no. 2 (1994): 29–64.

———. *Pandora's Hope: Essays in the Reality of Science Studies.* Cambridge, MA: Harvard University Press, 1999.

———. "Morality and Technology: The End of the Means." *Theory, Culture & Society* 19, no. 5–6 (2002): 247–260.

———. "What If We Talked Politics a Little?" *Contemporary Political Theory* 2, no. 2 (2003): 143–164.

———. "Why Has Critique Run out of Steam? From Matters of Fact to Matters of Concern." *Critical Inquiry* 30 (2004): 225–248.

Lautenschlager, Karl. "Controlling Military Technology." *Ethics* 95, no. 3 (1985): 692–711.

Lazarsfeld, Paul, and Robert Merton. "Mass Communication, Popular Taste and Organized Social Action." In *The Process and Effects of Mass Communication*, edited by Wilbur Schramm and Donald F. Roberts, revised ed., 554–578. Chicago: University of Illinois Press, 1971.

Leaksource. "NSA Chief Keith Alexander Keynote @ Black Hat USA 2013 (w/ Slide Presentation)." August 1, 2013. http://leaksource.info/2013/08/01/nsa-chief-keith-alexander-keynote-black-hat-usa-2013-w-slide-presentation/.

Lears, Jackson. *Something for Nothing: Luck in America*. New York: Viking, 2003.

Ledgett, Richard. "The NSA Responds to Edward Snowden's TED Talk." Filmed March 20, 2014, at TED 2014, Vancouver, British Columbia, Canada. Video, 33:19. http://www.ted.com/talks/richard_ledgett_the_nsa_responds_to_edward_snowden_s_ted_talk.

Lee, Micah. "Edward Snowden Explains How To Reclaim Your Privacy." *The Intercept*, November 12, 2015. https://theintercept.com/2015/11/12/edward-snowden-explains-how-to-reclaim-your-privacy/.

Leiter, Brian. "The Hermeneutics of Suspicion: Recovering Marx, Nietzsche, and Freud." In *The Future for Philosophy*, edited by Brian Leiter, 74–105. Oxford: Clarendon Press, 2004.

Lemov, Rebecca. *Database of Dreams: The Lost Quest to Catalog Humanity*. New Haven, CT: Yale University Press, 2015.

———. "Guantanamo's Catch-22: The Uncertain Interrogation Subject." In *Modes of Uncertainty—Anthropological Cases*, edited by Limor Samimian-Darash and Paul Rabinow, 88–104. Chicago: Chicago University Press, 2015.

———. "Archives-of-Self: The Vicissitudes of Time and Self in a Technologically Determinist Future." In *Science in the Archives: Pasts, Presents, Futures*, edited by Lorraine Daston, 247–270. Chicago: University of Chicago Press, 2017.

———. "Hawthorne's Renewal: Quantified Total Self." In *Humans and Machines at Work: Monitoring, Surveillance and Automation in Contemporary Capitalism*, edited by Phoebe V. Moore, Martin Upchurch, and Xanthe Whittaker, 181–202. London: Palgrave Macmillan, 2018.

Leopold, Jason. "Inside Washington's Quest to Bring Down Edward Snowden." *Vice News*, June 4, 2015. https://news.vice.com/article/exclusive-inside-washingtons-quest-to-bring-down-edward-snowden.

Lepore, Jill. "After the Fact." The *New Yorker*, March 21, 2016. http://www.newyorker.com/magazine/2016/03/21/the-internet-of-us-and-the-end-of-facts?mbid=social_facebook.

Leppakorpi, Lasse. "Beddit Presentation." Presentation given at MoneyTalks, Tampere, FL, February 10, 2011. Uploaded to SlideShare, February 14, 2011. https://www.slideshare.net/TechnopolisOnline/beddit-presentation.

Leys, Ruth. "The Turn to Affect: A Critique." *Critical Inquiry* 37, no. 3 (2011): 434–472.

Loewenstein, Antony. "The Ultimate Goal of the NSA Is Total Population Control." *The Guardian*, July 10, 2014. http://www.theguardian.com/commentisfree/2014/jul/11/the-ultimate-goal-of-the-nsa-is-total-population-control.

Lovece, Frank. "Soldier Showdown: Joe and Anthony Russo Take the Helm of 'Captain America' Franchise." *Filmjournal*, March 25, 2014. http://www.filmjournal.com/node/9232.

Lupton, Deborah. "Quantifying the Body: Monitoring and Measuring Health in the Age of MHealth Technologies." *Critical Public Health* 23, no. 4 (2013): 393–403.

———. "Understanding the Human Machine." *IEEE Technology and Society Magazine* 32, no. 4 (2013): 25–30.

———. "Self-Tracking Cultures: Towards a Sociology of Personal Informatics." Presentation at the OzCHI '14: Proceedings of the 26th Australian Computer-Human Interaction Conference: Designing Futures, the Future of Design, Sydney, December 2–5, 2014.

———. *The Quantified Self: A Sociology of Self-Tracking*. Cambridge: Polity Press, 2016.

———, Sarah Pink, Christine Heyes Labond, and Shanti Sumartojo. "Personal Data Contexts, Data Sense, and Self-Tracking Cycling." *International Journal of Communication* 12 (2018): 647–666.

Lyon, David. "Surveillance, Power and Everyday Life." In *Oxford Handbook of Information and Communication Technologies*, edited by Chrisanthi Avgerou, Robin Mansell, Danny Quah, and Roger Silverstone, 449–472. Oxford: Oxford University Press, 2007.

Maass, Peter. "Inside NSA, Officials Privately Criticize 'Collect It All' Surveillance." *The Intercept*, May 28, 2015. https://theintercept.com/2015/05/28/nsa-officials-privately-criticize-collect-it-all-surveillance/.

Mackenzie, Adrian. "The Production of Prediction: What Does Machine Learning Want?" *European Journal of Cultural Studies* 18, no. 4–5 (2015): 429–445.

Mackenzie, Donald. "Marx and the Machine." *Technology and Culture* 25, no. 3 (1984): 473–502.

———. "Unlocking the Language of Structured Securities." *Financial Times*, August 18, 2010. http://www.ft.com/cms/s/0/8127989a-aae3-11df-9e6b-00144feabdco.html#axzz3w2fTTGpL.

Magnani, Lorenzeo. *Morality in a Technological World: Knowledge as Duty*. Cambridge: Cambridge University Press, 2007.

Mailland, Julien, and Kevin Driscoll. *Minitel: Welcome to the Internet*. Cambridge, MA: MIT Press, 2017.

Majaca, Antonia. "Little Daniel Before the Law: Algorithmic Extimacy and the Rise of the Paranoid Apparatus." *E-Flux* 75 (2016). https://www.e-flux.com/journal/75/67140/little-daniel-before-the-law-algorithmic-extimacy-and-the-rise-of-the-paranoid-apparatus/.

Manjoo, Fahrad. "Mysteries of Sleep Lie Unsolved." *The New York Times*, February 24, 2015. http://www.nytimes.com/2015/02/26/technology/personaltech/despite-the-promise-of-technology-the-mysteries-of-sleep-lie-unsolved.html?_r=0.

Manokha, Ivan. "New Means of Workplace Surveillance: From the Gaze of the Supervisor to the Digitalization of Employees." *Monthly Review*, February 1, 2019. https://monthlyreview.org/2019/02/01/new-means-of-workplace-surveillance/.

Maris, Elena, Timothy Libert, and Jennifer Henrichsen. "Tracking Sex: The Implications of Widespread Sexual Data Leakage and Tracking on Porn Websites." *New Media & Society*, 2020. https://arxiv.org/abs/1907.06520.

Marvin, Carolyn. *When Old Technologies Were New: Thinking About Electric Communication in the Late Nineteenth Century*. Oxford: Oxford University Press, 1988.

Marx, Leo. "Technology: The Emergence of a Hazardous Concept." *Technology and Culture* 51, no. 3 (2010): 561–577.

———. "The Idea of 'Technology' and Postmodern Pessimism." In *Does Technology Drive History? The Dilemma of Technological Determinism*, edited by Merritt Roe Smith and Leo Marx, 237–258. Cambridge, MA: MIT Press, 1994.

Massumi, Brian. *Parables for the Virtual—Movement, Affect, Sensation*. Durham, NC: Duke University Press, 2002.

———. "Fear (The Spectrum Said)." *Positions* 13, no. 1 (2005): 31–48.

———. "Potential Politics and the Primacy of Preemption." *Theory & Event* 10, no. 2 (2007). https://muse.jhu.edu/issue/11693.

———. "The Future Birth of the Affective Fact: The Political Ontology of Threat." In *The Affect Theory Reader*, edited by Melissa Gregg and Gregory J. Seigworth, 52–70. Durham, NC: Duke University Press, 2010.

Matviyenko, Svitlana. "Interpassive User: Complicity and the Returns of Cybernetics." *The Fibreculture Journal*, no. 25 (2015): 135–163.

McCauley, Clark, and Sophia Moskalenko. "Toward a Profile of Lone Wolf Terrorists: What Moves an Individual from Radical Opinion to Radical Action." *Terrorism and Political Violence* 26, no. 1 (2014): 69–85.

McDonald, Aleecia, and Lorrie Faith Cranor. "The Cost of Reading Privacy Policies." *I/S—A Journal of Law and Policy for the Information Society* 4, no. 3 (2008): 1–22.

McLuhan, Marshall. *Understanding Media: The Extensions of Man*. New York: Mentor, 1964.

McQuillan, Dan. "Algorithmic States of Exception." *European Journal of Cultural Studies* 18, no. 4–5 (2015): 564–576.

Mearian, Lucas. "Insurance Company Now Offers Discounts—If You Let It Track Your Fitbit." *Computerworld*, April 17, 2015. http://www.computerworld.com/article/2911594/insurance-company-now-offers-discounts-if-you-let-it-track-your-fitbit.html.

Melley, Timothy. *Empire of Conspiracy: The Culture of Paranoia in Postwar America*. Ithaca, NY: Cornell University Press, 2000.

———. *The Covert Sphere: Secrecy, Fiction, and the National Security State*. Ithaca, NY: Cornell University Press, 2012.

Merleau-Ponty, Maurice. *Phenomenology of Perception*. Translated by Donald A Landes. London: Routledge, 2012.

Michael, George. *Lone Wolf Terror and the Rise of Leaderless Resistance*. Nashville, TN: Vanderbilt University Press, 2012.

Milburn, Tahl. "How My Life Automation System Quantifies My Life." Presentation at the Quantified Self 2015 Conference, San Francisco, June 18–20, 2015.

Milburne, Colin. "{Zero Day} // Hacking as Applied Science Fiction." Presentation at the Department of History and Sociology of Science Fall 2015 Monday Workshop Series, Philadelphia, 2015.

Milner, Michael. "Did Edward Snowden Tell Us Anything We Didn't Already Know?" *Chicago Reader*, June 25, 2013. http://www.chicagoreader.com/Bleader/archives/2013/06/25/did-edward-snowden-tell-us-anything-we-didnt-already-know.

Moeckli, Daniel. *Human Rights and Non-Discrimination in the "War on Terror."* Oxford: Oxford University Press, 2008.

Molotch, Harvey Luskin. *Against Security: How We Go Wrong at Airports, Subways, and Other Sites of Ambiguous Danger*. Princeton, NJ: Princeton University Press, 2012.

Monahan, Torin. *Surveillance in the Time of Insecurity*. New Brunswick, NJ: Rutgers University Press, 2010.

Moon, Il Chul, and Kathleen M. Carley. "Modeling and Simulating Terrorist Networks in Social and Geospatial Dimensions." *IEEE Intelligent Systems* 22, no. 5 (2007): 40–49.

Moor, James H. "The Nature, Importance, and Difficulty of Machine Ethics." In *Machine Ethics*, edited by Michael Anderson and Susan Leigh Anderson, 13–20. Cambridge: Cambridge University Press, 2011.

———. "What Is Computer Ethics?" *Metaphilosophy* 16, no. 4 (1985): 266–275.

Moore, Phoebe V. *The Quantified Self in Precarity—Work, Technology and What Counts*. Abingdon, UK: Routledge, 2018.

Morga, Alicia. "Do You Measure Up?" Fast Company, April 5, 2011. http://www.fastcompany.com/1744571/do-you-measure.

Morozov, Evgeny. *To Save Everything, Click Here: The Folly of Technological Solutionism*. New York: Public Affairs, 2013.

———. "The Taming of Tech Criticism." *The Baffler*, no. 27 (March 2015). https://thebaffler.com/salvos/taming-tech-criticism.

Morris, Robert R. "A Shocking Solution to Facebook Addiction." *Medium*, August 21, 2013. https://medium.com/@robertrmorris/a-shocking-solution-to-facebook-addiction-d1f5a14e2943.

Mortensen, Pete. "The Future of Technology Isn't Mobile, It's Contextual." *Fast Company*, May 24, 2013. http://www.fastcodesign.com/1672531/the-future-of-technology-isnt-mobile-its-contextual.

Morton, Timothy. *Hyperobjects: Philosophy and Ecology after the End of the World*. Minneapolis: University of Minnesota Press, 2013.

Muller, Benjamin J. "Securing the Political Imagination: Popular Culture, the Security Dispositif and the Biometric State." *Security Dialogue* 39, no. 2–3 (2008): 199–220.

Mumford, Lewis. *Technics and Civilization*. New York: Harcourt, Brace and World, 1963.

Munthe, Christian. *The Price of Precaution and the Ethics of Risk*. Dordrecht: Springer, 2011.

Nafus, Dawn, and Jamie Sherman. "This One Does Not Go Up to 11: The Quantified Self Movement as an Alternative Big Data Practice." *International Journal of Communication* 8 (2014): 1784–1794.

Nagel, Thomas. "Concealment and Exposure." *Philosophy & Public Affairs* 27, no. 1 (1998): 3–30.

Nakashima, Ellen. "NSA Cites Case as Success of Phone Data-Collection Program." *The Washington Post*, August 8, 2013. https://www.washingtonpost.com/world/national-security/nsa-cites-case-as-success-of-phone-data-collection-program/2013/08/08/fc915e5a-feda-11e2-96a8-d3b921c0924a_story.html.

Nakashima, Ellen, and Joby Warrick. "For NSA Chief, Terrorist Threat Drives Passion to 'Collect It All.'" *The Washington Post*, July 14, 2013. https://www.washingtonpost.com/world/national-security/for-nsa-chief-terrorist-threat-drives-passion-to-collect-it-all/2013/07/14/3d26ef80-ea49-11e2-a301-ea5a8116d211_story.html.

Nation, The. "Interview: Glenn Greenwald." 3 News (New Zealand), September 13, 2014.

National Commission on Terrorist Attacks upon the United States. "The 9/11 Commission Report: Final Report of the National Commission on Terrorist Attacks upon the United States: Executive Summary," 2004. http://govinfo.library.unt.edu/911/report/911Report_Exec.pdf.

National Security Archive. "Targeting Rationale." n.d. https://nsarchive2.gwu.edu/NSAEBB/NSAEBB436/docs/EBB-125.pdf.

Neff, Gina, and Dawn Nafus. *Self-Tracking*. Cambridge, MA: MIT Press, 2016.

Neuman, Bruce. "When a Senator and the C.I.A. Clash." *The New York Times*, March 12, 2014. https://www.nytimes.com/2014/03/13/opinion/when-a-senator-and-the-cia-clash.html.

Ngai, Sianne. *Ugly Feelings*. Cambridge, MA: Harvard University Press, 2005.

Nichols, Sam. "Your Phone Is Listening and It's Not Paranoia." *Vice Media*, June 4, 2018. https://www.vice.com/en_au/article/wjbzzy/your-phone-is-listening-and-its-not-paranoia.

Noble, Safiya Umoja. *Algorithms of Oppression: How Search Engines Reinforce Racism*. New York: New York University Press, 2018.

Nørretranders, Tor. *The User Illusion—Cutting Consciousness Down to Size*. New York: Viking, 1998.

North, Anna. "Why You Want an App to Measure Calories but Not Character." *The Washington Post*, August 26, 2014. http://op-talk.blogs.nytimes.com/2014/08/26/why-you-want-an-app-to-measure-calories-but-not-character/?_r=1.

Nye, David E. *American Technological Sublime*. Cambridge, MA: MIT Press, 1994.

Nyong'o, Tanya. "Plenary 4." Presentation at the Affect Theory Conference | Worldings | Tensions | Futures in Lancaster, PA, October 14–17, 2015.

Obama, Barack. "Memorandum for the Heads of Executive Departments and Agencies: Transparency and Open Government." The White House, January 21, 2009. https://www.whitehouse.gov/the_press_office/TransparencyandOpenGovernment.

———. "Transcript: Obama's Remarks on NSA Controversy." *The Wall Street Journal*, March 14, 2013. http://blogs.wsj.com/washwire/2013/06/07/transcript-what-obama-said-on-nsa-controversy/.

Oboler, Andre, Kristopher Welsh, and Lito Cruz. "The Danger of Big Data: Social Media as Computational Social Science." *First Monday* 17, no. 7 (2012). https://firstmonday.org/article/view/3993/3269/.

O'Harrow Jr., Robert. "NSA Chief Asks a Skeptical Crowd of Hackers to Help Agency Do Its Job." *The Washington Post*, July 31, 2013. https://www.washingtonpost.com/world/national-security/nsa-chief-asks-a-skeptical-crowd-of-hackers-to-help-agency-do-its-job/2013/07/31/351096e4-fa15-11e2-8752-b41d7ed1f685_story.html.

Oliver, John, Tim Carvell, James Taylor, and Jon Thoday. Last Week Tonight with John Oliver. Aired April 5, 2015, on HBO.

Olson, Parmy. "Fitbit Data Now Being Used in the Courtroom." *Forbes*, November 16, 2014. http://www.forbes.com/sites/parmyolson/2014/11/16/fitbit-data-court-room-personal-injury-claim.

O'Malley, Pat. *Risk, Uncertainty and Government*. London: Glasshouse Press, 2004.

O'Neill, Cathy. *Weapons of Math Destruction: How Big Data Increases Inequality and Threatens Democracy*. New York: Crown, 2016.

Oreskes, Naomi, and Erik M. Conway. *Merchants of Doubt: How a Handful of Scientists Obscured the Truth on Issues from Tobacco Smoke to Global Warming*. New York: Bloomsbury Press, 2010.

Packer, George. "The Holder of Secrets." *The New Yorker*, October 20, 2014. http://www.newyorker.com/magazine/2014/10/20/holder-secrets.

Paglen, Trevor. "Invisible Images (Your Pictures Are Looking at You)." *The New Inquiry*, December 8, 2016. https://thenewinquiry.com/invisible-images-your-pictures-are-looking-at-you/.

Pantucci, Raffaello. *A Typology of Lone Wolves: Preliminary Analysis of Lone Islamist Terrorists*. London: The International Centre for the Study of Radicalisation and Political Violence, 2011. http://mediafieldsjournal.squarespace.com/rise-of-the-imsi-catcher/.

Papoulias, Constantina, and Felicity Callard. "Biology's Gift: Interrogating the Turn to Affect." *Body & Society* 16, no. 1 (2010): 29–56.

Parikka, Jussi. "Operative Media Archaeology: Wolfgang Ernst's Materialist Media Diagrammatics." *Theory, Culture & Society* 28, no. 5 (2011): 52–74.

Parks, Lisa. "Rise of the IMSI Catcher." *Media Fields* 11 (2016).

Pasquale, Frank. *The Black Box Society: The Secret Algorithms that Control Money and Information*. Cambridge, MA: Harvard University Press, 2015.

———."Machine Learning, Meaning, & Law." Presentation at the Cybernetics Conference, New York, November 15–17, 2017.

———. "Odd Numbers." *Real Life*, August 20, 2018. http://reallifemag.com/odd-numbers/.

Patterson, John. "How Hollywood Softened Us up for NSA Surveillance." *The Guardian*, June 16, 2013. http://www.theguardian.com/film/shortcuts/2013/jun/16/hollywood-softened-us-up-nsa-surveillance.

Pearson, Erika. "Smart Objects, Quantified Selves, and a Sideways Flow of Data." Presentation at ICA 2016, Fukuoka, Japan, June 9–13, 2016.

Peters, John Durham. *The Marvelous Clouds: Toward a Philosophy of Elemental Media.* Chicago: University of Chicago Press, 2015.

Pew Research Center. "Majority Views NSA Phone Tracking as Acceptable Anti-Terror Tactic," June 10, 2013.

Pfaller, Robert. "Interpassivity and Misdemeanors. The Analysis of Ideology and the Zizekian Toolbox." *International Journal of Zizek Studies* 1, no. 1 (2001): 33–50.

———. "Little Gestures of Disappearance(1) Interpassivity and the Theory of Ritual." *Journal of European Psychoanalysis* 16 (2003). www.journal-psychoanalysis.eu/little-gestures-of-disappearance1-interpassivity-and-the-theory-of-ritual/.

Pincus, Walter. "Questions for Snowden." *The Washington Post*, July 8, 2013. https://www.washingtonpost.com/world/national-security/questions-for-snowden/2013/07/08/d06ee0f8-e428-11e2-80eb-3145e2994a55_story.html.

Poindexter, John. "Information Awareness Office Overview." Presentation at DARPATech 2002, Anaheim, CA, August 2, 2002.

"Policy Strategy No.5: Reclaiming the Public Spaces of New York." New York, 1994.

Porter, Theodore M. *Trust in Numbers: The Pursuit of Objectivity in Science and Public Life.* Princeton, NJ: Princeton University Press, 1995. https://www.ncjrs.gov/pdf-files1/Photocopy/167807NCJRS.pdf.

———. "The Objective Self." *Victorian Studies* 50, no. 4 (2014): 641–647.

"Pplkpr." *pplkpr.com.* Accessed March 31, 2016. http://www.pplkpr.com.

Procter, Robert N. "Agnotology: A Missing Term to Describe the Cultural Production of Ignorance (and Its Study)." In *Agnotology: The Making & Unmaking of Ignorance,* edited by Robert N. Procter and Londa Schiebinger, 1–33. Stanford, CA: Stanford University Press, 2008.

Proud, James. "Goodbye, Hello." *Medium*, June 12, 2017. https://medium.com/@hello/goodbye-hello-c62ea1f58d13.

Puar, Jasbir K., and Amit S. Rai. "Monster, Terrorist, Fag: The War on Terrorism and the Production of Docile Patriots." *Social Text* 20, no. 3 (2002): 117–148.

Public Policy Polling. "Walker Leads Tightly Clustered GOP Field, Clinton Up Big Nationally." May 13, 2015. https://www.publicpolicypolling.com/wp-content/uploads/2017/09/PPP_Release_National_51315.pdf.

Rabouin, David. "Styles in Mathematical Practice." In *Cultures Without Culturalism: The Making of Scientific Knowledge,* edited by Karine Chemla and Evelyn Fox Keller, 196–223. Durham, NC: Duke University Press, 2017.

Ramirez, Ernesto. "Larry Smarr: Where There Is Data There Is Hope." *Quantified Self*, February 12, 2013. http://quantifiedself.com/2013/02/larry_smarr_croneshope_in_data/.

———. "QS Access: Personal Data Freedom." *Quantified Self*, February 11, 2015. http://quantifiedself.com/2015/02/qs-access-personal-data-freedom/.

Rappaport, Roy. *Ritual and Religion in the Making of Humanity.* Cambridge: Cambridge University Press, 1999.

Reagle, Joseph. *Hacking Life: Systematized Living and Its Discontents*. Cambridge, MA: MIT Press, 2019.

Reeves, Danny. "Frictionless Tracking with Beeminder Autodata." Presentation at the Quantified Self 2015 Conference, San Francisco, June 18–20, 2015.

Reeves, Joshua. *Citizen Spies: The Long Rise of America's Surveillance Society*. New York: New York University Press, 2017.

Reitman, Janet. "Jahar's World." *Rolling Stone*, July 17, 2013. http://www.rollingstone.com/culture/news/jahars-world-20130717.

Rettberg, Jill Walker. *Seeing Ourselves Through Technology: How We Use Selfies, Blogs and Wearable Devices to See and Shape Ourselves*. Basingstoke, UK: Palgrave Macmillan, 2014.

Rheingold, Howard. *The Virtual Community: Homesteading on the Electronic Frontier*. Cambridge, MA: MIT Press, 2000.

Rich, Steven, and Matt DeLong. "NSA Slideshow on 'The TOR Problem.'" *The Washington Post*, October 4, 2013. http://apps.washingtonpost.com/g/page/world/nsa-slideshow-on-the-tor-problem/499/.

Rieder, Bernhard. "Scrutinizing an Algorithmic Technique: The Bayes Classifier as Interested Reading of Reality." *Information, Communication & Society* 20, no. 1 (2017): 100–117.

Risen, James, and Laura Poitras. "N.S.A. Report Outlined Goals for More Power." *The New York Times*, November 22, 2013. http://www.nytimes.com/2013/11/23/us/politics/nsa-report-outlined-goals-for-more-power.html.

Rose, Nikolas. "At Risk of Madness." In *Embracing Risk: The Changing Culture of Insurance and Responsibility*, edited by Tom Baker and Jonathan Simon, 209–237. Chicago: Chicago University Press, 2002.

———, Pat O'Malley, and Mariana Valverde. "Governmentality." *Annual Review of Law and Social Science* 2, no. 1 (2006): 83–104.

Rosen, David, and Aaron Santesso. *The Watchman in Pieces—Surveillance, Literature, and Liberal Personhood*. New Haven, CT: Yale University Press, 2013.

Rosenberg, Daniel. "Data before the Fact." In *Raw Data Is an Oxymoron*, edited by Lisa Gitelman, 15–40. Cambridge, MA: MIT Press, 2013.

Rothman, Joshua. "'Person of Interest': The TV Show that Predicted Edward Snowden." *The New Yorker*, January 14, 2014. http://www.newyorker.com/culture/culture-desk/person-of-interest-the-tv-show-that-predicted-edward-snowden.

Rotman, Brian. *Becoming Beside Ourselves: The Alphabet, Ghosts, and Distributed Human Being*. Durham, NC: Duke University Press, 2008.

Rouvroy, Antoinette. "Privacy, Data Protection, and the Unprecedented Challenges of Ambient Intelligence." *Studies in Ethics, Law, and Technology* 2, no. 1 (2008): 1–51.

Rowan, David. "Snowden: Big Revelations to Come, Reporting Them Is Not a Crime." *Wired*, March 18, 2014. http://www.wired.co.uk/news/archive/2014-03/18/snowden-ted.

Ruckenstein, Minna, and Mika Pantzar. "Beyond the Quantified Self: Thematic Exploration of a Dataistic Paradigm." *New Media & Society* 119, no. 3(2017): 401-418.

Ruffino, Paolo. "Games to Live With." *Digital Culture & Society* 2, no. 1 (2016): 153–160.

Ruppert, Evelyn. "Population Objects: Interpassive Subjects." *Sociology* 45, no. 2 (2011): 218–233.

Rusbridger, Alan. "The Snowden Leaks and the Public." *The New York Review of Books*, November 21, 2013. http://www.nybooks.com/articles/2013/11/21/snowden-leaks-and-public/.

Ryan, Susan Elizabeth. *Garments of Paradise: Wearable Discourse in the Digital Age.* Cambridge, MA: MIT Press, 2014.

Samimian-Darash, Limor. "Governing Future Potential Biothreats." *Current Anthropology* 54, no. 1 (2013): 1–22.

———, and Paul Rabinow. "Introduction." In *Modes of Uncertainty—Anthropological Cases*, edited by Limor Samimian-Darash and Paul Rabinow, 1–9. Chicago: Chicago University Press, 2015.

Sampson, Tony D. *The Assemblage Brain: Sense Making in Neuroculture.* Minneapolis: University of Minnesota Press, 2017.

Saury, Jean-Michel. "The Phenomenology of Negation." *Phenomenology and the Cognitive Sciences* 8, no. 2 (2008): 245–260.

Schick, Camilla, and Stephen Castle. "'I Trusted Him': London Attacker Was Friendly with Neighbors." *The New York Times*, June 5, 2017. https://www.nytimes.com/2017/06/05/world/europe/london-attack-theresa-may.html?action=click&contentCollection=Europe&module=RelatedCoverage&region=EndOfArticle&pgtype=article.

Schneier, Bruce. "It's Smart Politics to Exaggerate Terrorist Threats." CNN, May 20, 2013. https://www.cnn.com/2013/05/20/opinion/schneier-security-politics/index.html.

———. *Data and Goliath—The Hidden Battles to Collect Your Data and Control Your World.* New York: W. W. Norton, 2015.

Schölzel, Hagen. "Beyond Interactivity. The Interpassive Hypotheses on 'Good Life' and Communication." Presentation at ICA 2014, Seattle, May 22–26, 2014.

———. "Backing Away from Circles of Control: A Re-Reading of Interpassivity Theory's Perspectives on the Current Political Culture of Participation." *Empedocles: European Journal for the Philosophy of Communication* 8, no. 2 (2017): 187–203.

Schudson, Michael. *The Rise of the Right to Know—Politics and the Culture of Transparency, 1945–1975.* Cambridge, MA: Harvard University Press, 2015.

Schüll, Natasha Dow. "Data for Life: Wearable Technology and the Design of Self-Care." *Biosocieties* 11, no. 3 (2016): 317–333.

———. "Self-Tracking Technology from Compass to Thermostat." Presentation at Streams of Consciousness conference, University of Warwick, Warwick, UK, April 21–22, 2016.

Schulte, Joachim. "World-Picture and Mythology." *Inquiry: An Interdisciplinary Journal of Philosophy* 31 (1988): 323–334.

Schwartz, Matthias. "The Whole Haystack." *The New Yorker*, January 26, 2015. http://www.newyorker.com/magazine/2015/01/26/whole-haystack.

Scott, James C. *Seeing like a State: How Certain Schemes to Improve the Human Condition Have Failed.* New Haven, CT: Yale University Press, 1998.

Sekula, Allan. "The Body and the Archive." *October* 39 (1986): 3–64.

Selinger, Evan. "Why It's OK to Let Apps Make You a Better Person." *The Atlantic*, March 9, 2012. https://www.theatlantic.com/technology/archive/2012/03/why-its-ok-to-let-apps-make-you-a-better-person/254246/.

sen.se. "Mother." N.d. Accessed March 29, 2016. https://sen.se/mother.

Sethi, Maneesh. "Why I Hired a Girl on Craigslist to Slap Me in the Face—And How It Quadrupled My Productivity." *Hack the System*, October 16, 2012. http://hackthesystem.com/blog/why-i-hired-a-girl-on-craigslist-to-slap-me-in-the-face-and-why-it-quadrupled-my-productivity/.

Shapin, Steven. *The Scientific Life: A Moral History of a Late Modern Vocation.* Chicago: University of Chicago Press, 2008.

Sharon, Tamar, and Dorien Zandbergen. "From Data Fetishism to Quantifying Selves: Self-Tracking Practices and the Other Values of Data." *New Media & Society* 19, no. 11 (2017): 1695–1709.

Shaver, Andrew. "You're More Likely to Be Fatally Crushed by Furniture than Killed by a Terrorist." *The Washington Post*, November 23, 2015. https://www.washingtonpost.com/news/monkey-cage/wp/2015/11/23/youre-more-likely-to-be-fatally-crushed-by-furniture-than-killed-by-a-terrorist/.

Shin, Seung back, and Kim Yong hun. *Cloud Face.* Shinseungback Kimyonghun, 2012. http://ssbkyh.com/works/cloud_face/.

———. *Flower.* Shinseungback Kimyonghun, 2017. http://ssbkyh.com/works/flower/.

Shorrock, Tim. "US Intelligence Is More Privatized than Ever Before." *The Nation*, September 16, 2015. http://www.thenation.com/article/us-intelligence-is-more-privatized-than-ever-before/.

Siebers, Tobin. *Cold War Criticism and the Politics of Skepticism.* New York: Oxford University Press, 1993.

Siegel, Barry. "Judging State Secrets: Who Decides—and How?" In *After Snowden: Privacy, Secrecy, and Security in the Information Age*, edited by Ronald Goldfarb, 141–190. New York: St. Martin's Press, 2015.

Sifton, John. "A Brief History of Drones." *The Nation*, February 7, 2012. http://www.thenation.com/article/brief-history-drones/.

Simon, Jeffrey D. *Lone Wolf Terrorism: Understanding the Growing Threat.* New York: Prometheus Books, 2013.

Singer, Peter W. "Military Robots and the Future of War." *The New Atlantis* 23 (2009): 25–45.

———. *Wired for War: The Robotics Revolution and Conflict in the 21st Century.* New York: Penguin, 2009.

———. "The Ethics of Killer Applications: Why Is It So Hard to Talk about Morality When It Comes to New Military Technology?" *Journal of Military Ethics* 9, no. 4 (2010): 299–312.

Skiba, Russell J. "Zero Tolerance, Zero Evidence: An Analysis of School Disciplinary Practice." Research Policy Report #SR2. Bloomington: Indiana Education Policy Center, 2000.

Sloterdijk, Peter. *Critique of Cynical Reason*. Translated by Michael Eldred. Minneapolis: University of Minnesota Press, 1987.

Smith, Daniel W. "Deleuze and the Question of Desire: Toward an Immanent Theory of Ethics." *Parrhesia* 2 (2007): 66–78.

Smith, Michael L. "Recourse of Empire: Landscapes of Progress in Technological America." In *Does Technology Drive History?: The Dilemma of Technological Determinism*, edited by Merritt Roe Smith and Leo Marx, 37–52. Cambridge, MA: MIT Press, 1994.

Smith, Neil. "Global Social Cleansing: Postliberal Revanchism and the Export of Zero Tolerance." *Social Justice* 28, no. 3 (2001): 68–74.

Snowden, Edward. "Whistleblower Edward Snowden Gives 2013's Alternative Christmas Message." *Channel4.com*, December 25, 2013. http://www.channel4.com/programmes/alternative-christmas-message/on-demand/58816-001.

Sollie, Paul. "Ethics, Technology Development and Uncertainty: An Outline for Any Future Ethics of Technology." *Journal of Information, Communication and Ethics in Society* 5, no. 4 (2007): 293–306.

Spaaij, Ramon. *Understanding Lone Wolf Terrorism: Global Patterns, Motivations and Prevention*. Dordrecht: Springer, 2012.

Spiekermann, Sarah, and Frank Pallas. "Technology Paternalism—Wider Implications of Ubiquitous Computing." *Poiesis Und Praxis* 4, no. 1 (2006): 6–18.

Springer, Nathan R. "Patterns of Radicalisation: Identifying the Markers and Warning Signs of Domestic Lone Wolf Terrorists in Our Midst." Thesis, Naval Postgraduate School, 2009.

Srnicek, Nick. *Platform Capitalism*. Cambridge: Polity Press, 2017.

Stalcup, Meg. "Policing Uncertainty: On Suspicious Activity Reporting." In *Modes of Uncertainty—Anthropological Cases*, edited by Limor Samimian-Darash and Paul Rabinow, 69–87. Chicago: Chicago University Press, 2015.

Star, Susan Leigh, and James R. Griesemer. "Institutional Ecology, `Translations' and Boundary Objects: Amateurs and Professionals in Berkeley's Museum of Vertebrate Zoology, 1907–39." *Social Studies of Science* 19, no. 3 (1989): 387–420.

Steedman, Carolyn. *Dust: The Archive and Cultural History*. New Brunswick, NJ: Rutgers University Press, 2002.

Sterling, Bruce. "The Internet of Things: Quantified Self, IoT, Smart Cities, Smart Cars, Smart Clothes." *Wired*, April 3, 2013. http://www.wired.com/2013/04/the-internet-of-things-quantified-self-iot-smart-cities-smart-cars-smart-clothes/.

Sterman, David. "Infographic: How the Government Exaggerated the Successes of NSA Surveillance." *Slate*, January 16, 2014. http://www.slate.com/blogs/future_tense/2014/01/16/nsa_surveillance_how_the_government_exaggerated_the_way_its_programs_stopped.html.

Stevenson, David. "'The Gudeman of Ballangeich': Rambles in the Afterlife of James V." *Folklore* 115, no. 2 (2004): 187–200.

Steyerl, Hito. "A Sea of Data: Apophenia and Pattern (Mis-) Recognition." *E-Flux* 72 (2016). https://www.e-flux.com/journal/72/60480/a-sea-of-data-apophenia-and-pattern-mis-recognition/.

Stinson, Liz. "Having a Hard Time Being a Human? This App Manages Friendships for You." *Wired*, January 26, 2015. http://www.wired.com/2015/01/hard-time-human-app-manages-friendships/.

Stout, Matt, and Donna Goodison. "Dzhokhar Tsarnaev Loves Pot, Wrestling Say Friends." *Boston Herald*, April 20, 2013. http://www.bostonherald.com/news_opinion/local_coverage/2013/04/dzhokhar_tsarnaev_loves_pot_wrestling_say_friends.

Swan, Melanie. "The Quantified Self: Fundamental Disruption in Big Data Science and Biological Discovery." *Big Data* 1, no. 2 (2013): 85–99.

Szpunar, Piotr M. "From the Other to the Double: Identity in Conflict and the Boston Marathon Bombing." *Communication, Culture & Critique* 9, no. 4 (2015): 577–594.

———. "Premediating Predisposition: Informants, Entrapment, and Connectivity in Counterterrorism." *Critical Studies in Media Communication* 34, no. 4 (2017): 371–385.

Tan, Ling. "Reality Mediators," November 22, 2015. http://lingql.com/reality-mediators/.

Tarnoff, Ben. "The Data Is Ours!" *Logic* 4 (2018): 91–108.

Taussig, Michael. *Defacement—Public Secrecy and the Labour of the Negative*. Stanford, CA: Stanford University Press, 1999.

Taylor, Emmeline. *Surveillance Schools: Security, Discipline and Control in Contemporary Education*. Basingstoke, UK: Palgrave Macmillan, 2013.

Thaler, Richard H., and Cass R. Sunstein. *Nudge: Improving Decisions about Health, Wealth and Happiness*. New York: Penguin Books, 2009.

Thatcher, Jim, David O'Sullivan, and Dillon Mahmoudi. "Data Colonialism through Accumulation by Dispossession: New Metaphors for Daily Data." *Environment and Planning D: Society and Space* 34, no. 6 (2016): 990–1006.

Thomas, Suzanne L., Dawn Nafus, and Jamie Sherman. "Algorithms as Fetish: Faith and Possibility in Algorithmic Work." *Big Data & Society* 5, no. 1 (2018).

Thompson, E. P. "Time, Work-Discipline, and Industrial Capitalism." *Past and Present* 38 (1967): 56–97.

Thrift, Nigel. "Lifeworld Inc—and What to Do about It." *Environment and Planning D: Society and Space* 29, no. 1 (2011): 5–26.

Turner, Fred. *From Counterculture to Cyberculture: Stewart Brand, the Whole Earth Network, and the Rise of Digital Utopianism*. Chicago: University of Chicago Press, 2006.

Turner, Victor. "Liminal to Liminoid." In *Play, Flow, Ritual: An Essay in Comparative Symbology*, edited by Victor Turner, 53–92. New York: Performing Arts Journal Publishing, 1982.

Upchurch, Martin, and Phoebe V. Moore. "Deep Automation and the World of Work." In *Humans and Machines at Work: Monitoring, Surveillance and Automation in Contemporary Capitalism*, edited by Phoebe V Moore, Martin Upchurch, and Xanthe Whittaker, 45–71. London: Palgrave Macmillan, 2018.

Urist, Jacoba. "From Paint to Pixels." *The Atlantic*, May 14, 2015. http://www.theatlantic.com/entertainment/archive/2015/05/the-rise-of-the-data-artist/392399/.

US Department of Defense Education Activity. "Operations Security (OPSEC)." Accessed March 18, 2016. www.dodea.edu.

US Department of Justice. "Two Illinois Men Charged With Conspiring to Provide Material Support to ISIS." April 12, 2017. https://www.justice.gov/opa/pr/two-illinois-men-charged-conspiring-provide-material-support-isis..

US Department of State. "1996 Patterns of Global Terrorism Report," 1997. https://www.fbi.gov/file-repository/stats-services-publications-terror_96.pdf/view.

———. "Patterns of Global Terrorism, 1995," 1996. https://fas.org/irp/threat/terror_95/index.html.

———. "Patterns of Global Terrorism 1999," 2000. https://www.fbi.gov/stats-services/publications/terror_99.pdf.

US Senate Select Committee on Intelligence. "Annual Open Hearing on Current and Projected National Security Threats to the United States." 113th Congress, Second Session, Washington, DC, January 29, 2014. http://www.intelligence.senate.gov/hearings/open-hearing-current-and-projected-national-security-threats-against-united-states#.

Vaidhyanathan, Siva. *The Googlization of Everything: (And Why We Should Worry)*. Los Angeles: University of California Press, 2011.

Vallor, Shannon. *Technology and the Virtues: A Philosophical Guide to a Future Worth Wanting*. Oxford: Oxford University Press, 2016.

Valverde, Mariana. *Diseases of the Will: Alcohol and the Dilemmas of Freedom*. Cambridge: Cambridge University Press, 1998.

Van Buren, Peter. "10 Myths about NSA Surveillance that Need Debunking." *MotherJones*, January 13, 2013. http://www.motherjones.com/politics/2014/01/10-myths-nsa-surveillance-debunk-edward-snowden-spying.

van Dijck, José. *The Culture of Connectivity: A Critical History of Social Media*. Cambridge: Oxford University Press, 2013.

van Doorn, John. "An Intimidating New Class: The Physical Elite." *The New York Magazine*, May 29, 1978. http://nymag.com/news/features/49241/.

van Lente, Harro. "Forceful Futures: From Promise to Requirement." In *Contested Futures: A Sociology of Prospective Techno-Science*, edited by Nik Brown, Brian Rappert, and Andrew Webster, 43–64. Aldershot, UK: Ashgate, 2000.

van Oenen, Gijs. "A Machine that Would Go of Itself: Interpassivity and Its Impact on Political Life." *Theory & Event* 9, no. 2 (2006). https://muse.jhu.edu/article/198813.

———. "Interpassivity Revisited: A Critical and Historical Reappraisal of Interpassive Phenomena." *International Journal of Zizek Studies* 2, no. 2 (2002). https://www.zizekstudies.org/index.php/IJZS/article/view/80/0.

Vattimo, Gianni. *The Transparent Society*. Baltimore: Johns Hopkins University Press, 1992.

Verdugo, Richard R. "Race-Ethnicity, Social Class, and Zero-Tolerance Policies." *Education and Urban Society* 35, no. 1 (2002): 50–75.

Vickers, Debbie, prod., and Jay Leno. *The Tonight Show with Jay Leno*. New York: NBC, August 6, 2013.

Viseu, Ana, and Lucy Suchman. "Wearable Augmentations." In *Technologized Images, Technologized Bodies: Anthropological Approaches to a New Politics of Vision*, edited by Jeanette Edwards, Penelope Harvey, and Peter Wade, 161–84. New York: Berghahn Books, 2010.

Wadhwa, Vivek. "Five Innovation Predictions for 2013." *The Washington Post*, January 4, 2013. https://www.washingtonpost.com/national/on-innovations/five-innovation-predictions-for-2013/2013/01/04/f4718be6-55c5-11e2-bf3e-76c0a789346f_story.html.

Wainwright, Oliver. "Rise of the 'Inner-Net': Meet the Most Connected Man on the Planet." *The Guardian*, March 19, 2014. https://www.theguardian.com/artanddesign/architecture-design-blog/2014/mar/19/inner-net-most-connected-man-earth-fitness-trackers-data.

Wajcman, Judy, and Emily Rose. "Constant Connectivity: Rethinking Interruptions at Work." *Organization Studies* 32, no. 7 (2011): 941–961.

Waldman, Peter, Lizette Chapman, and Jordan Robertson. "Palantir Knows Everything About You." *Bloomberg Businessweek*, April 19, 2018. https://www.bloomberg.com/features/2018-palantir-peter-thiel/.

Walker, Rob. "Wasted Data." *The New York Times*, December 3, 2010. http://www.nytimes.com/2010/12/05/magazine/05FOB-Consumed-t.html?mtrref=undefined&gwh=224196DF50B6FEC7F871890CABC54F5A&gwt=pay.

Wallace, Lane. "The Illusion of Control." *The Atlantic*, May 26, 2010. http://www.theatlantic.com/technology/archive/2010/05/the-illusion-of-control/57294/.

Warschauer, Mark, and Morgan Ames. "Can One Laptop Per Child Save the World's Poor?" *Journal of International Affairs* 64, no. 1 (2010): 33–51.

The Washington Post. "Transcript: Senate Intelligence Hearing on National Security Threats." January 29, 2014. https://www.washingtonpost.com/world/national-security/transcript-senate-intelligence-hearing-on-national-security-threats/2014/01/29/b5913184-8912-11e3-833c-33098f9e5267_story.html.

Watson, Sara M. "Living with Data: Personal Data Uses of the Quantified Self." Master's thesis. University of Oxford, 2013.

———. "You Are Your Data." *Slate*, November 12, 2013. www.slate.com/articles/technology/future_tense/2013/11/quantified_self_self_tracking_data_we_need_a_right_to_use_it.html.

———. "Data Doppelgängers and the Uncanny Valley of Personalization." The *Atlantic*, June 16, 2014. https://www.theatlantic.com/technology/archive/2014/06/data-doppelgangers-and-the-uncanny-valley-of-personalization/372780/.

———. "Data Is the New '___.'" *dis magazine*, 2015. http://dismagazine.com/blog/73298/sara-m-watson-metaphors-of-big-data/.

Weinberger, David. "Our Machines Now Have Knowledge We'll Never Understand." *Wired*, April 18, 2017. https://www.wired.com/story/our-machines-now-have-knowledge-well-never-understand/.

Weiser, Mark. "The Computer for the 21st Century." *Scientific American* (September 1991): 94–104.

Weizenbaum, Joseph. *Computer Power and Human Reason: From Judgment to Calculation*. New York: W. H. Freeman, 1976.

Wilcox, Lauren. "Embodying Algorithmic War: Gender, Race, and the Posthuman in Drone Warfare." *Security Dialogue* 48, no. 1 (2016): 11–28.

Williams, Matt. "Edward Snowden Is a 'Traitor' and Possible Spy for China—Dick Cheney." *The Guardian*, June 16, 2013. https://www.theguardian.com/world/2013/jun/16/nsa-whistleblower-edward-snowden-traitor-cheney.

Williams, Rosalind H. *The Triumph of Human Empire: Verne, Morris, and Stevenson at the End of the World*. Chicago: University of Chicago Press, 2013.

Wingfield, Nick. "Gauging the Natural, and Digital, Rhythms of Life." *The New York Times*, June 19, 2013. http://bits.blogs.nytimes.com/2013/06/19/gauging-the-natural-and-digital-rhythms-of-life/?_r=0.

Winner, Langdon. *Autonomous Technology: Technics-out-of-Control as a Theme in Political Thought*. Cambridge, MA: MIT Press, 1977.

———. "Do Artifacts Have Politics?" *Daedalus* 109, no. 1 (1980): 121–136.

Wired. "The Decades that Invented the Future, Part 12: The Present and Beyond." February 8, 2013. https://www.wired.com/2013/02/the-decades-that-invented-the-future-part-12-the-present-and-beyond/.

Wittes, Benjamin. "The Problem at the Heart of the NSA Disputes: Legal Density." *Lawfare*, February 14, 2014. https://www.lawfareblog.com/problem-heart-nsa-disputes-legal-density.

Wittgenstein, Ludwig. *On Certainty*. Edited by G. E. M Anscombe and G. H von Wright. New York: Harper Torchbooks, 1969.

Wolf, Gary. "Know Thyself: Tracking Every Facet of Life, from Sleep to Mood to Pain, 24/7/365." *Wired*, June 22, 2009. http://www.wired.com/2009/06/lbnp-knowthyself/.

———. "The Data-Driven Life." *The New York Times*, April 28, 2010. http://www.nytimes.com/2010/05/02/magazine/02self-measurement-t.html?_r=1.

———. "Tim Ferriss Wants to Hack Your Body." *Wired*, November 29, 2010. http://www.wired.com/2010/11/mf_qa_ferriss/.

———. "What Is the Quantified Self?" *Quantified Self*, March 3, 2011. http://quantifiedself.com/2011/03/what-is-the-quantified-self/.

———. "A Public Infrastructure for Data Access." *Quantified Self*, March 8, 2016. http://quantifiedself.com/2016/03/larry-smarr-interview/.

———, and Kevin Kelly. "Wired's Gary Wolf & Kevin Kelly Talk the Quantified Self." Presentation at the WIRED Health Conference: Living by Numbers, New York, October 15–16, 2012.

Woods, Chris. "The Story of America's Very First Drone Strike." The *Atlantic*, May 30, 2015. http://www.theatlantic.com/international/archive/2015/05/america-first-drone-strike-afghanistan/394463/.

Zaret, David. *Origins of Democratic Culture: Printing, Petitions, and the Public Sphere in Early-Modern England*. Princeton, NJ: Princeton University Press, 2000.

Zemrani, Laila. "Using Self Tracking to Exercise More Efficiently." Presentation at the New York Quantified Self Meetup, New York, 2015.

Zetter, Kim. "Snowden Smuggled Documents From NSA on a Thumb Drive." *Wired*, June 13, 2013. http://www.wired.com/2013/06/snowden-thumb-drive/.

Zielinski, Siegfried. *Deep Time of the Media: Toward an Archaeology of Hearing and Seeing by Technical Means*. Cambridge, MA: MIT Press, 2006.

Zimmer, Michael. "The Externalities of Search 2.0: The Emerging Privacy Threats When the Drive for the Perfect Search Engine Meets Web 2.0." *First Monday* 13, no. 3 (2008). http://firstmonday.org/article/view/2136/1944.

Zimmerman, Jess. "There's a Fitness Tracker for Vaginas. Quantifying Your Life Has Gone Too Far." The *Guardian*, July 14, 2014. http://www.theguardian.com/commentisfree/2014/jul/14/fitness-tracker-vagina-quantified-life.

Žižek, Slavoj. "The Inherent Transgression." *Cultural Values* 2, no. 1 (1998): 1–17.

———. "The Interpassive Subject," n.d. http://www.egs.edu/faculty/slavoj-zizek/articles/the-interpassive-subject/.

Zuboff, Shoshana. "Big Other: Surveillance Capitalism and the Prospects of an Information Civilization." *Journal of Information Technology* 30, no. 1 (2015): 75–89.

———. *The Age of Surveillance Capitalism: The Fight for a Human Future at the New Frontier of Power*. New York: PublicAffairs, 2019.

Zylinska, Joanna. *Nonhuman Photography*. Cambridge, MA: MIT Press, 2017.

ABOUT THE AUTHOR

Sun-ha Hong analyzes the fantasies, values, and sentimentalities surrounding big data and artificial intelligence. By mapping out the historical and philosophical roots of how we think about objectivity and progress, he seeks to help clarify the moral stakes of technological promises. Hong's research includes the contested legacy of the Enlightenment in contemporary technoculture, the sociopolitical implications of imagined technofutures, and the mediated history of the public sphere. Hong is currently Assistant Professor of Communication at Simon Fraser University. He was previously a Mellon Postdoctoral Fellow at MIT, and has a PhD from the University of Pennsylvania.

www.ingramcontent.com/pod-product-compliance
Lightning Source LLC
Chambersburg PA
CBHW020247030426
42336CB00010B/660

INDEX

absent presence, 56, 128

actionism, 153, 154

alethurgy, 82. *See also* Foucault, Michel

Alexander, Keith, 41, 59, 126, 145

Amazon, 38, 101, 102, 113

apophenia, 39, 41

Beddit (sensor), 13, 80

better knowledge, 1, 2, 3, 7–12, 14, 16, 19, 21, 22, 24, 27, 56, 135, 157, 182, 187, 188, 195, 198

Big Brother, 27, 75, 86, 176. *See also* Orwell, George

Borges, Jorge Luis, 33, 53; Borgesian, 178. *See also* Library of Babel

"broken windows," 143. *See also* zero tolerance

Butler, Judith, 71, 229

Clapper, James, 116, 117, 119, 121, 128, 149

Cold War effect, 41. *See also* paranoia

conjectural knowledge, 78. *See also* Ginzburg, Carlo

connect the dots, 39, 55, 58, 136; "all the dots," 59; connecting the dots, 38, 60, 134; connect more dots, 74

control creep, 102, 175, 185

Dancy, Chris, 105–6, 162

Daston, Lorraine, 16–17, 156, 172. *See also* Galison, Peter

data hunger, 54, 58, 59, 60, 63, 109; data hungry, 62, 65, 73

"data is the new oil," 175, 194

data market, 8, 12, 25–26, 28, 77, 89, 100, 104, 108, 113, 178, 185, 189. *See also* surveillance capitalism

data-sense, 9, 156–57, 160–62, 164, 165, 166, 168, 169, 170, 171, 172, 173, 174, 175, 181, 182, 184, 197, 240

data's intimacy, 79, 83, 91, 95, 109

Dean, Jodi, 123. *See also* public supposed to believe (PSB); public supposed to know (PSK).

deepfakes, 158–59

Deleuze, Gilles, 1, 75, 177

Derrida, Jacques, 35, 117

digital hygiene, 122, 139

disinfectant, 37, 55; to disinfect, 48. *See also* purity

dragnet, 134, 203; dragnet surveillance, 5, 33, 58, 127, 132, 182, 194, 237

drone, 145–46, 155

Enlightenment, 9, 16, 17, 28, 31, 43, 46, 57, 78, 169, 175, 188, 209

exoself, 164, 173. *See also* Kelly, Kevin

exposed body, 98. *See also* hijacked body

fabrication, 2–4, 8, 16, 20, 24, 49, 73, 78, 80, 95, 110, 115, 140, 145, 146, 151, 154, 156, 157, 158, 173, 179, 181–82; data-driven fabrication, 10, 161; duality of fabrication, 9, 77; preemptive fabrication, 137, 140, 143

Facebook, 38, 95, 100, 101, 122, 168

facial recognition, 183, 193, 197

FBI, 20, 66, 68, 116, 137–39